THE

ELEGANCE OF RUTHLESSNESS

HOW TO DEFEAT ISIS AND REGAIN U.S. DOMINANCE IN MILITARY EVOLUTION

BY

THOMAS ST. JAMES

The Elegance of Ruthlessness: How to Defeat ISIS and Regain U.S. Dominance in Military Evolution

This book is a work of fiction. Names, characters, and incidents are products of the author's imagination or are used fictitiously. Any resemblance to persons living or dead is entirely coincidental. Geographic villages, rivers, and valleys are, however, actual locations in Afghanistan. Maps were produced though GoogleEarth, ©2014 Cnset/Spot; Image 2014 © DigitalGlobe.

Library of Congress Cataloging-in-Publication Data, Copyright 2014 Thomas St. James

"The Elegance of Ruthlessness"
ISBN-10: 0-9905-5632-8
ISBN-13:978-0-9905-5632-9

DEDICATION

This book is dedicated in memory of Chris Kyle, Navy SEAL, America's best sniper ever, and a modern hero of the great State of Texas.

Forward

"What are we doing here?!?!" Mary* was the task force targeting analyst. She was brilliant and extremely experienced. She had just stormed into my office with her hair on fire and flames jumping out of her eyes. "Really, what are we doing here?!?!" We were in Eastern Afghanistan. It was 2011 – America's tenth year at war in that country. She and I had served in this theater before. It was my fourth tour and her third. She exclaimed her question both out of frustration and in a rhetorical form. "Do you know what I've been doing all day? I've been watching Mohammedullah* [the leader for the Taliban in all of Eastern Afghanistan] and Omar al-Masri* [the al Qaida Emir for all of Eastern Afghanistan] sitting on top of a roof chatting together all afternoon – they were having a meeting together!!" Translation: She had been watching the two most dangerous terrorist leaders in all of Eastern Afghanistan- one an Afghan, the other an Arab- from the video feed of a powerful camera mounted on a very expensive unmanned aerial vehicle (UAV). She was watching them plot the killing of American and Afghan soldiers…at the least. She continued. "And WHAT did we do about it!? WHAT?! NOTHING!! We sat there and watched. Our command wouldn't fire on them because they were on top of a building and we created a rule that we cannot destroy buildings in Afghanistan!! Again I ask you – what are we doing here?!" I didn't want to provide the obvious answer. The answer was – nothing. We weren't doing anything and therefore we were losing. We were losing a war to a group of men who lived in the 13th century. Why were we losing? Answering that last question is only a small part of what I hope to answer in this book.

The evolution of warfare is primarily being driven in the East, not the West. The drivers of the evolution of warfare are stateless groups like al Qaida, the "Islamic State", and Hezbollah, through their use of "unconventional warfare," and in states like China and Russia, the foremost leaders in cyber-warfare *and* in using *economic levers* as strategic weapons. Calling terrorism or cyber-warfare "illegal" or "against the internationally recognized rules of warfare" doesn't make them any less effective. They are, in fact, very effective means of executing warfare because unenforced (or unenforceable) rules are, *de facto,* not rules at all.

The purpose of this book is not to criticize a certain political party's administration or a particular general or admiral or group of experienced military brass. My purpose is broader. I hope to demonstrate, through logic, a fictional anecdotal demonstration, and history, that to succeed in war, one must be a leader in the evolution of war. The United States is the greatest nation that is and ever has been in the world. One of our greatest strengths is our ability to innovate. It is part of our national DNA. Yet, today, the US is not leading the evolution or innovation of warfare.

America has a choice. It is my hope that we will see the errors of our ways and once again regain the mantle as the world's greatest innovators at warfare so that we can defend our way of life and live a more peaceful and prosperous existence. TSJ - 2012

*NOTE: Names above have been changed for operational security reasons.

<u>Table of Contents</u>

PART I

Never Ending Battle

"I am your enemy therefore I am your teacher."
- Mazar Rakham – Ender's Game

"There is only victory & defeat. The winner gets to take the moral high-ground because they get to write the history books. The loser just loses...."

- H from Unthinkable

Introduction to "The Elegance of Ruthlessness"

My name is Captain Jack B.F. Jennings and this is my story. It is a most extraordinary story. Some would almost call it weird. This is the story of a portion of my experiences as an infantry officer fighting the Taliban in the Watapur Valley, Konar Province, Afghanistan. Now, if you're reading this book, you must be interested in military history and tactics and such. In fact, you're probably a member of the military, more than likely, in the Army or Marine Corps. Since you're probably in the military, you probably noticed my middle initials are "B.F." Sad but true. I won't belabor the profane version of what this usually stands for in the military. You know and I know, it doesn't really mean "Blue Falcon." I was burdened with these two initials by no choice of my own. It is an old family name. The initials go back to my great-grandfather on my mother's side, Lieutenant Backsight Forethought, or, as the British pronounce it, "Leftenant" Backsight Forethought. For, you see, my great-grandfather was in the British Army. He gained some notoriety during the Boer War fighting guerrillas in a locale known as Duffer's Drift. After that war, he fought with distinction in World War I, the "war to end all wars." It was fascinating to read his journals I inherited as he described the dramatic differences between fighting irregular Boer guerrillas compared to fighting regular German soldiers ten years later. Following the Great War, he immigrated to the United States. His son, my grandfather, Hindsight Forethought, fought in World War II. His nephew, my father, Jimmy B.F. Jennings, faced combat in the waning years of the Vietnam War. He was the first to acquire the "B.F." initials, which, I unfortunately inherited. And so, my family, and the wars we have fought, have come full circle: from combatting unconventional Boers, to battling regular German soldiers (twice), to facing a mixture of Vietcong irregulars and North Vietnamese conventional troops to my fight against (highly) unconventional Islamic guerrillas.

It is unfortunate that I did not read my great-grandfather's journals until after I had completed my tour in Afghanistan. Nor did I read my grandfather's. Heck, my father never even kept a journal. But, like most young men in their 20's, I really didn't give a hoot about history. At the time, I was far more interested in chasing girls and having fun with my friends in the army, that is, when I wasn't training in the field or on the battlefield itself. I also wish I had studied history more. It might have prevented me from getting into the queer predicament that I found myself amid the rugged mountains of the Hindu Kush in eastern Afghanistan. I also wish I had maintained a journal. I did not. Who writes journals in the era of Facebook and Twitter? But, now I have instead written a book. And now, I shall tell you my story, because, if I didn't tell you my story, then I truly would be a "Blue Falcon."

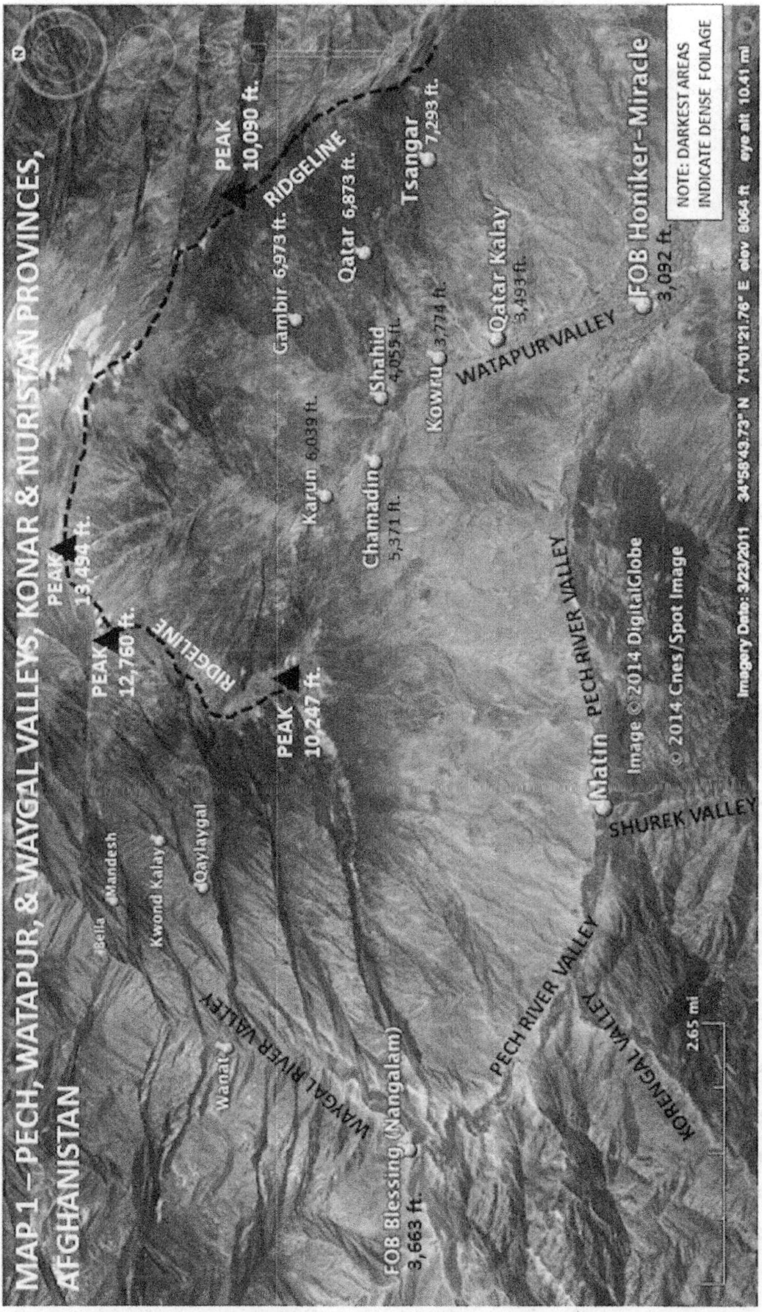
MAP 1 – PECH, WATAPUR, & WAYGAL VALLEYS, KONAR & NURISTAN PROVINCES, AFGHANISTAN
PEAK 10,090 ft.
RIDGELINE
PEAK 13,494 ft.
PEAK 12,760 ft.
RIDGELINE
PEAK 10,247 ft.
Gambir 6,973 ft.
Qatar 6,873 ft.
Tsangar 7,293 ft.
Karun 6,039 ft.
Shahid 4,055 ft.
Kowru 3,774 ft.
Qatar Kalay 3,493 ft.
Chamadin 5,371 ft.
WATAPUR VALLEY
FOB Honiker–Miracle 3,092 ft.
NOTE: DARKEST AREAS INDICATE DENSE FOILAGE
Bella
Mandesh
Kwond Kalay
Qaylaygal
Wanat
WAYGAL RIVER VALLEY
FOB Blessing (Nangalam) 3,663 ft.
PECH RIVER VALLEY
PECH RIVER VALLEY
Matin
SHUREK VALLEY
KORENGAL VALLEY
Image © 2014 DigitalGlobe
© 2014 Cnes/Spot Image
2.65 mi
Imagery Date: 3/23/2011 34°58'43.73" N 71°01'21.76" E elev 8064 ft eye alt 10.41 mi

CHAPTER 1 - The Shura

My eyes groggily flittered open. My head pounded and my ears were ringing. I moved my arm slowly and noticed a tug at my skin. Looking over, I saw an IV with a large gauge needle sticking in my arm.

"Where am I?" I muttered to the uniformed woman intently studying charts at my side.

"You are at the FOB (Forward Operating Base) Wright FAS - Forward Aid Station," she replied. "Do you know why you're here?"

I thought for a moment as glimpses of memories began to drift into my head. The last thing I remembered I was headed to a shura at FOB Blessing deep in the Pech River Valley. Now I was at the FAS in Asadabad.

"I think...I think we were hit by an IED. We were in a gunfight."

"That's right," she said. "What else can you remember?"

Straining past the sledge hammers pounding at my head, it all started coming back to me.

The day had began like any other day. Local Afghan civic leaders from a variety of key villages asked to hold a shura, a meeting between their principal leaders and ours. My commander, LTC Lemoux, was at FOB Blessing while I was at FOB Honiker-Miracle, which sat at the mouth of the Watapur Valley east of FOB Blessing. The shura leaders asked to hold the shura at the Monogai District Center, a few hundred yards from FOB Blessing, because it was more centrally located in the Konar Province. Travel was too difficult for them, they complained. None of them had cars. They were forced to take taxis, walk, or ride in donkey carts. They asked to have the meeting at 1300 local hours (1PM), following noon prayers and well before the evening, sunset prayer. Any other time, they lamented, would interfere with their religious practices. So, LTC Lemoux agreed and called all of his company commanders in to attend. Two of the company commanders had to fly in by helicopter to Blessing, one from FOB Bostic and the other from FOB Joyce. Since my FOB was closest to Blessing and the District Center, and helicopters were in limited supply, I planned to drive to the shura in a convoy of MRAP (Mine-Resistant Ambush Protected) armored cars.

Traveling in convoys in MRAPs was something we did routinely. At the beginning of my tour, I'd rehearse the convoy procedures after submitting a 5 paragraph operations order (OPORD) every time. But, after a while, and after being in a few TICs (Troops in Contact - e.g. getting shot at) all my men knew the drill and I didn't have the time for detailed rehearsals. I did, however, ensure that all my NCOs conducted detailed PCCs (Pre-Combat Checks) and PMCS (Preventative Maintenance Checks and Services) on all the MRAP and weapons prior to venturing outside the wire. (I learned in Iraq that complacency got men killed.)

We departed the FOB at 0800 sharp in a convoy of six vehicles, four from one of my platoons, my MRAP, and one other. At the FOB, I left my second in command, executive officer 1LT Mike Mitchell, behind along with two of my platoons and my mortar section, and my artillery slice. The ride from Honiker-Miracle to the Monogai District Center was uneventful. Most of the guys took their mind off of the boredom of the two hour convoy by listening to heavy metal music and smoking cigarettes. For them, this was just another "milk-run."

The shura also turned out to be uneventful. The long bearded local leaders, the ends of their gray beards dyed orange with henna, started out polite. Then, like every other shura I had been to, the meeting digressed into a bitch session. "*We want more wells dug for us; we want more blankets from the UNHCR; we want more free food from USAID, we want a solatia payment for a woman your mortars accidentally killed...blah, blah, blah.*" After my first six months in Afghanistan, I never really understood why FM 3-24, the US Army Counterinsurgency Manual, put such emphasis on shuras and KLEs (Key Leader Engagements). It sounded like good doctrine when I was at FT Benning. Lord knows, it was written by generals with far more experience than me. It was all about winning the "*hearts and minds*" of the locals. But, the reality was, it never seemed to get us anywhere. The KLEs seemed like mushy feel-good meetings; like something a high-school counselor (us) would conduct for troubled kids in a school full of gangs (them). Perhaps it made some of the local Afghan leaders feel good about us. But, one could never tell from their stone faced bearded expressions. I mean, really, how can one *see* inside the *heart and mind* of a man?

Following the shura, we mounted up and got ready to roll. I was glad to see my men remembered to re-load rounds in the chambers of their crew-served weapons. We were always careful to clear our weapons immediately upon entering a FOB or a urban area with civilians like the District Center (located in Nangalam Village). There had been a few times on return trips when my men had complacently forgotten to re-load their weapons during return trips. But, my NCO's put a stop to that!

I climbed into my MRAP and mashed the button to close the heavy steel door. The hydraulics hissed as they sealed the door. MRAPs were so comfortable. They were a million times better than the up-armored HMMWV's I had endured in Iraq as a young lieutenant. They felt like a climate-controlled spaceship. I buckled into my 4-point harness and allowed the comfort of the air conditioning to waft over my sweat laced body, made all the more hot by the suffocating body armor I was wearing. I was a turtle in a spaceship guarded by a mighty .50 caliber M2 "Maw-duce" machine gun. We started to roll east down the nicely paved road, paid for courtesy of US tax-payers. I peeked out of one of the windows. My gunners were up and scanning attentively. I felt invincible.

Suddenly, I saw a blinding flash followed by a massive explosion. Even with the muffling of my radio headset, the thunderous explosion made my ears ring. Our lead vehicle had been hit by an improvised explosive device (IED). "*Contact, contact!*" one of the MRAP commanders shouted over the radio. A cacophony of the . 50 cals erupted from nearby turret gunners letting off a stream of deadly bullets. I strained to look around through the windows of my vehicle. They seemed to be firing at rocks a mere 20 feet from the road. My gunner, Rodriguez, joined in the fire.

"Rod, what are you shooting at?" I shouted.

"Hell, sir, they're right there...right there on the side of the road!"

From his higher vantage point, Rod could see the enemy dismounts where I could not. I heard a drumming at the side of my MRAP. The Taliban were firing 7.62 rounds into the side of my vehicle. A few hit the windows, causing the glass to spiderweb, but not shatter. Now I was blind. I ordered the number two vehicle forward to support the lead MRAP, whose two front tires had been blown off. Depressing the mike, I called higher for CAS (Close Air Support). Taliban RPGs (rocket propelled grenades) began slamming into the sides of our MRAP. Fortunately, most detonated prior to contact by the protective cage surrounding our MRAP. Higher told me A-10 Warthogs jets were on station and headed to our location. I started to wonder why our electronic jammers hadn't detonated the IED before Vehicle 1 crossed the bomb's location.

BLAAAAM!!!

A massive explosion rocked my MRAP. That answered my question. The Taliban were using command detonated IEDs, not electronic or pressure plate detonated bombs. Our electronic jammers had no effect against a man hidden behind a rock that we could not see. The vehicle's V-shaped hull deflected the blast, but shredded our rear tires and shredding off the rear axle. We were now at a precarious angle and I spied Rodriguez struggling to aim with weight of the turret and the .50 cal, now canted from the loss of our rear tires. More RPGs hissed in the air and skipped along the pavement. The Taliban were now aiming at the tires of Vehicle 2 and 6 in an attempt to cripple them.

I cursed at the thought of the KLE. Someone from the shura must have tipped off the enemy about our convoy's timing and disposition. Bastards!

Again I shouted over the radio again for CAS. I gave the accurate grid coordinates of the enemy positions. Word came back directly from one of the pilots. He couldn't drop his bombs or fire his 40mm chain gun because the Taliban were right on top of us alongside the road - danger close. A volley of RPGs crashed on top of our MRAP. The enemy had the high-ground, and was looking for the soft spot at the top. A massive blast hit the turret. A spray of pink mist filled the MRAP and Rod's headless body slumped to the floor.

Another volley of RPGs must have slammed into our MRAP, because I completely blacked out. When I came to, I found myself in the FAS. But, in the fog of my pounding skull, important lessons floated through my brain. I drifted off again, determined to not make the same mistakes.

Lessons Learned:

- *Never let local nationals dictate the time and place to meet them. They need to answer your dictates. You're the boss - act like it!*

- *KLEs - Key Leader Engagements - can be a good tool for deception, but they will not dissuade your enemy from attacking you.*
- *Never forget that KLEs with local nationals have a direct line of communication to the people trying to kill you. Whether they are nefarious spies or not, they will tell their Jihadist friends EVERYTHING - including when and where your convoy is headed. [Hence, see the emphasis on "deception" above – they lie to you so you should lie to them.]*
- *"Trust but verify" is for politicians. Don't trust the bastards. Be a warrior.*
- *Local national "Leaders" aren't really leaders; the leaders of the insurgents are the leaders because they have the guns.*
- *The enemy knows what "danger-close" is and knows how to use it against you. Don't count on CAS from the USAF.*
- *Consider integrating a local leader (without his cell phone or radio) into the convoy somewhere. You're not using him as a "human shield" - you're just "protecting him along the route." (Okay - you really are using him as a human shield - so what, they do it all the time.)*

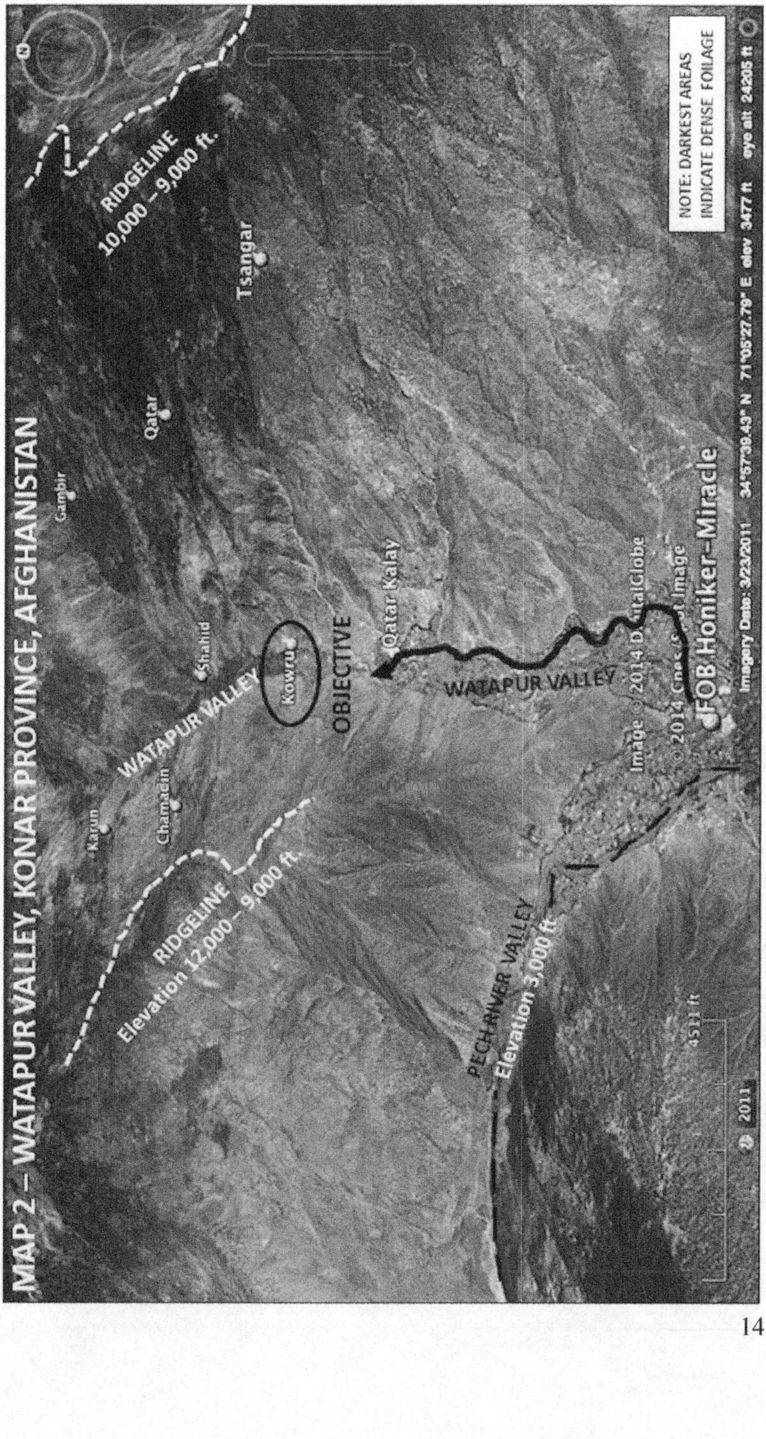
MAP 2 – WATAPUR VALLEY, KONAR PROVINCE, AFGHANISTAN
RIDGELINE
10,000 – 9,000 ft.
Tsangar
Qatar
Gambir
Shahid
Kowru
OBJECTIVE
Qatar Kalay
WATAPUR VALLEY
WATAPUR VALLEY
Karun
Charmadin
RIDGELINE
Elevation 12,000 – 9,000 ft.
PECH RIVER VALLEY
Elevation 3,000 ft.
FOB Honiker–Miracle
NOTE: DARKEST AREAS
INDICATE DENSE FOILAGE
Imagery Date: 3/23/2011 34°57'39.43" N 71°05'27.79" E elev 3477 ft eye alt 24205 ft
4511 ft
2011

CHAPTER 2 - Waltzing up the Watapur

I opened my eyes and to my surprise (and relief) I found myself back in my rack at FOB Honiker-Miracle. The IV was no longer in my arm. My mind was clear and now filled with a host of new lessons learned from the previous night's horrible dream.

Skipping breakfast, I called LTC Lemoux on the secure line and provided a suggestion: instead of gathering all the battalion's company commanders in one location and gathering all the regional village leaders for a massive shura, it might be better for us to hold our own local shuras with our own local, esteemed, Afghan village and district leaders. This would make for less moving parts and prevent us from putting "all our eggs in one basket."

"That's a great idea, Captain Jennings...I'm glad I thought of it..." replied my plucky commander.

So, I made an announcement to our district governor and told him to invite the local village leaders to meet for a local shura on my own FOB Honiker-Miracle. Thus, I would be keeping with the tenets of the counter-insurgency manual, still aiming to win over their hearts and minds to the wisdom of US foreign policy. Having a meeting with the local leaders would calculate into the political equation of my terrorist problem. I also learned at the Captain's Career Course the necessity of having local presence patrols. A public demonstration of US military might was necessary to assure those innocent civilians of

Watapur District that we cared about protecting them from the ravages of the terrorists in their midst.

Fresh with the memories of my previous night's nightmare, I knew that a little subterfuge would aid me in my planned presence patrol. I could not determine the exact loyalties of the local mullahs (or the governor, for that matter) but there was no need for me to telegraph how I was going to do business!

We set the time of the shura on my FOB for 1300, again aware of cultural sensitivities regarding Islamic prayer. I even ordered up some local flat-bread and slaughtered a goat to demonstrate my cultural sensitivities. They finally arrived (most of them about 1-2 hours late) and we ended up having a rather jovial banter about local wells, more free blankets from the UNHCR, and the possibility of building a new medical clinic (the previous clinic had been blown up by the Taliban the day after its opening ceremony, for even medical clinics provided with "infidel" funds were considered "haram.") The Afghan elders appeared impressed with my selection of foods, as well as the peaceful internal surroundings and strength of the base itself. All in all, nothing was decided or accomplished at this KLE, save one important thing: I told the local leaders that I was planning to have a presence patrol down the Pech River Road to inspect the damage done to the bridge next to the village of Matin in two days. I am proud to say, it was one of my most eloquent lies.

Two days after the shura, at approximately 0600, I was informed by my Tactical Operations Center (TOC) that an Afghan National Army (ANA) patrol headed west on the Pech River road had received a blistering attack near the village of Matin.

"Ah-ha!" I proclaimed to myself. "My ruse worked like magic." Someone who attended the shura indeed had tipped off the Taliban. The Taliban had assumed the ANA column was the advanced party of our forces. The insurgents must have shifted a large contingent of their terrorist forces to that location. Now was my opportunity to launch my presence patrol.

At the odd hour of 1015L, I sent out my second platoon on a presence patrol. I attentively stayed at the FOB to monitor the situation over the radio. However, instead of turning west down the Pech river to Matin, the platoon headed straight north, up the dirt road into the middle of the Watapur Valley itself. Their objective, as we rehearsed the prior day, was to drive north into the village of Kowru, dismount, inspect the local schools, and have an impromptu KLE with the elder mullah of the Kowru. Kowru would be the farthest north up the valley we had ever patrolled, even though it was only 2.5 miles from the FOB.

This presence patrol would also operate within the strict confines of the rules of engagement (ROE), which stipulated we could not fire unless fired upon first. So, in a sense, our patrol would act as live American bait. Anyone taking pop-shots at our MRAPs with AK-47's would incur the wrath of .50 cal machine guns, mortars, artillery, and one last trick I had up my sleeve - two AH-64 Apache helicopters who could fire with pin-point accuracy.

The MRAPs rumbled northward up the rocky road. We had been instructed by International Security Assistance Force headquarters (ISAF HQ), located in distant Kabul, to make every effort to wave to the people and look friendly. I found this silly, but many of the drivers and gunners tried to wave at the few pedestrians on the side of the road. The children and teenagers always looked so innocent and happy and usually waved back. These children were tech savvy, and the hope for the country's future. Many of them had taught their fathers and grandfathers how to use cell phones. Women, however, were nowhere to be seen. They were either in the wheat fields harvesting or at home in seclusion.

Second platoon continued calling in checkpoints along the route. Prior to the mission, with the help of Google Earth, we had pre-plotted the checkpoints, and had accurate grids saved in all the GPSs.

About one kilometer south of Kowru, the lead vehicle stopped as the gunner spied something suspicious on the road. It looked like an ordinary pile of rocks, but experience suggested it could be a command detonated IED.

"All stop, all stop," commanded the platoon leader.

I could not hear the internal banter on the platoon net, but was glad they were scanning for possible IED locations. As the platoon leader read off the ten-digit grid coordinates of the possible IED location via radio, I heard a massive explosion over his intercom.

"White 1, White 1, what was that...SITREP!" I instinctively shouted over the intercom.

It was not an IED. It was a salvo of RPG's focused on the rear vehicle, all slamming into it simultaneously. It was time to get off the "X" (i.e. out of the danger area) fast! The lead MRAP sprayed .50 cal rounds into the rocks where he believed an IED was buried. There was no explosion. The rocks placed on the road had merely been a ruse! The Taliban wanted the convoy to pause just long enough for their RPG gunners to take aim at the stationary rear vehicle. Their deception worked. Then all hell broke loose.

Fortunately, the RPG gunners' first salvo was aimed at center mass of the rear vehicle and the MRAP's protective cage had prematurely detonated the rockets. The rear MRAP's gunner took quick aim at the insurgents and hosed them down with .50 caliber rounds. All of our turret gunners focused on the mountainside to their right as panicky drivers threw transmissions into reverse. The Apaches swooped into action, scanning for armed combatants. A second salvo of RPGs hissed through the air, now focused on the tires of the rear vehicle. The two right tires of the rear MRAP were taken out, crippling the steel beast which now listed precariously to the side. The platoon leader ordered his vehicles to turn around, but not simultaneously. Two trucks kept their guns up and trained on the hillside, spraying insurgents with deadly fire, whenever they exposed themselves. The turning vehicles, however, had to make slow four-point turns because the precarious dirt road was so narrow. Exposing their rear, two trucks were pummeled by small arms fire (SAF) and RPGs. Whenever an insurgent exposed himself, however, an Apaches made mincemeat of him. Another truck took SAF to a tire, but the inflation system keep it pumped up enough to not be crippled. Most of the insurgents appeared to be about 100 feet above the MRAPs in the mountainside, cloaked by boulders and trees. Then they took aim with 7.62 rounds at the gunners and driver's windows. The Apaches the remained focused on the 100 foot elevation level, where the attackers were massed and hosed them down. The bad guys may have had the high-ground over my platoon, but, at 400 feet, the Apaches had the high-ground over them.

Or most of them. Hidden directly at the 400 foot mark on the side of the mountain were two DSHK-a anti-aircraft guns with deadly 12.7mm machine guns. Shouting "Allahu Akbar!" these gunners simultaneously opened up taking aim at the nearest Apache. Huge holes riddled its tail. One round caught the tail rotor, crippling it. Like a wounded bird, the Apache slowly descended in a controlled spiraling motion, auto-rotating down, down, down to a nearby field of wheat. At least the downed Apache's wing-man smoked both DSHK-a positions so they wouldn't be a threat again.

Listening to the action by radio, it was hard for me to fully discern heads or tails of the battle, but I knew my men were in trouble. Worse yet, now they had to fight their way to the downed helo pilots and conduct an extraction. Confusion swarmed over the platoon as one truck rushed to the aid of the helicopter. Half-way to the chopper, the heavy armored truck became mired in the muddy field. Now there were three crippled war machines in strewn in the middle of the hostile Watapur Valley: the platoon's rear vehicle (which blocked their egress route south), the Apache (in the middle of the wheat field), and one MRAP, (stuck between the road and the downed Apache). So much for a simple "presence patrol!"

Thankfully, I had planned a Quick Reaction Force (QRF) consisting of First Platoon. That platoon was on RECON 1 status, meaning, their trucks were running and the platoon's men were all geared up and ready to roll to the rescue. I was about to give the order to launch the QRF and…

BOOM!BOOOOM! FWAA-BOOOOOM BOOM BOOM!

I was hoping it was just thunder from the unpredictable Konar sky. It wasn't thunder. It was a barrage of mortar rounds landing right on top of my FOB! My TOC shook like an earthquake. I threw on my body army and rushed outside to get my bearings. As I exited, I saw a salvo of old Soviet mortars land square in the middle of my mortar positions inside the FOB! Anger and frustration engulfed me. The next explosion was spectacular. I saw the monumental flash slightly before I heard the explosion. The concussion lifted me ten feet through the air. My land kicked the wind out of me. Dazed and confused, I watched a huge yellow and black mushroom cloud appear where enemy mortars had just made a direct hit on the FOB's ASP (Ammunition Supply Point)! On the ground and covered with dust, I wondered, "How...how in the world did they fire mortars with such precision?" Just before passing out, the grinning image village elder with the rotting teeth I'd met at the shura on the FOB two days earlier formed in my mind and my tragic mistake became completely clear.

<u>Lessons Learned</u>

- *TV pundits and politicians often claim the need for a "political equation" to the Islamic terrorist problem. They chant "warfare alone is not a solution." Poppy-cock!! They never read Clausewitz:* "warfare is an <u>extension</u> of politics!"

- *That old man you met at the shura on your FOB...yea, he just drew a sketch of your fortifications and gave it to the enemy. (If you have to do a KLE, do it on terrain you can defend, that is not of the enemy's choosing, and not where he can spy on the inside of your fortifications!)*
- *Structuring a military operation to support a KLE is stupid. You should structure a KLE to support a military action.*
- *Those teenage kids you just passed by on the road waving at you with a cell phone in their pocket...yea, they just called their dad & uncle and told them you're coming. (Baby skunks don't grow up to be pet rabbits.)*
- *MRAPs are great at surviving bombs in the road, but not for leading an assault.*
- *Helicopter gunships are an awesome force multiplier, can dominate the high-ground, and are deadly accurate. But, as David slew Goliath, they can also be shot down by cavemen.*

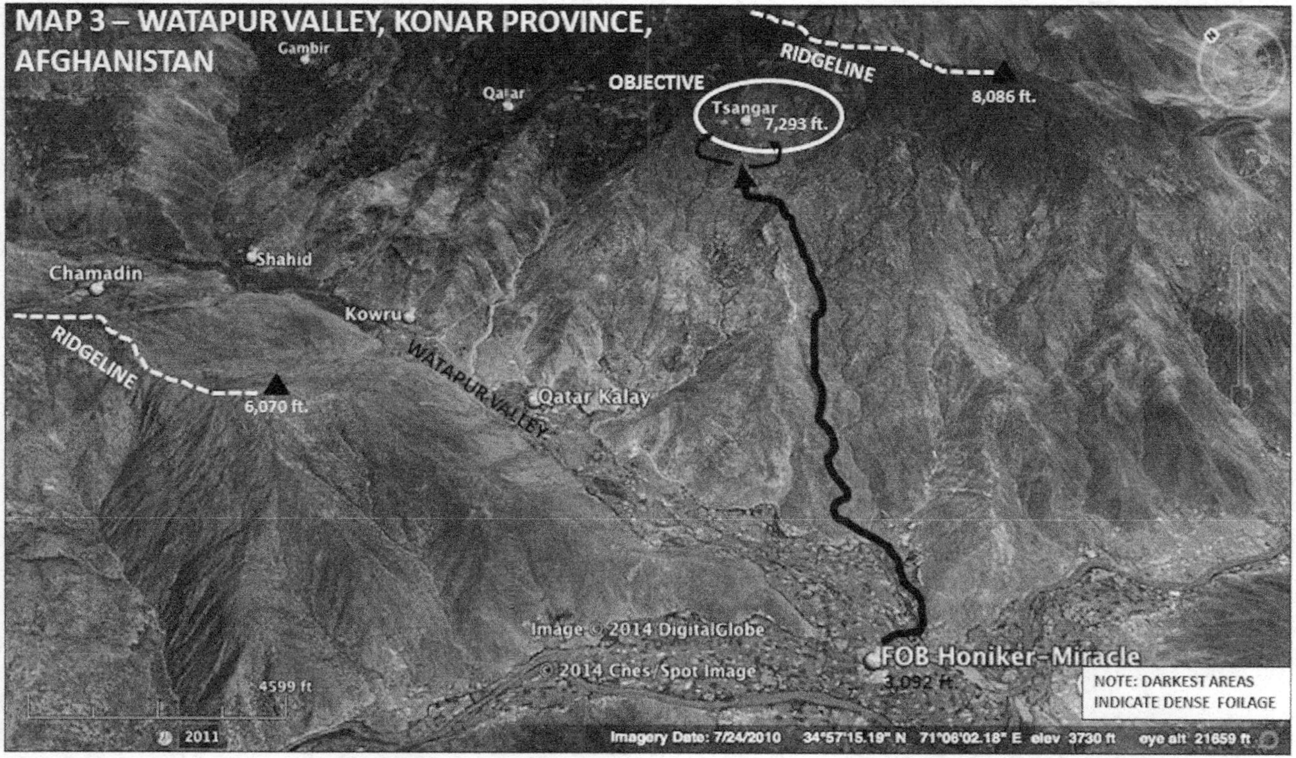
MAP 3 – WATAPUR VALLEY, KONAR PROVINCE, AFGHANISTAN
Gambir
Qatar
OBJECTIVE
Tsangar
7,293 ft.
RIDGELINE
8,086 ft.
Shahid
Chamadin
Kowru
RIDGELINE
WATAPUR VALLEY
6,070 ft.
Qatar Kalay
Image © 2014 DigitalGlobe
© 2014 Cnes/Spot Image
FOB Honiker-Miracle
3,092 ft.
NOTE: DARKEST AREAS
INDICATE DENSE FOILAGE
4599 ft
2011
Imagery Date: 7/24/2010 34°57'15.19" N 71°06'02.18" E elev 3730 ft eye alt 21659 ft

CHAPTER 3 - Charging Uphill

I woke up in a pool of goo. "*Oh my God,*" I panicked, "*my brains are pouring out.*" Slowly, I lifted my hand to feel for the entry wound. There was none. The darn A/C in my hooch went out again in the night; the goo was a pool of my own sweat mingled with the ever-present Afghan dust. The vivid image of the erupting ASP had been nothing more than a horrible vision in my sleep.

Despite the lack of cool air, I felt refreshed, possibly because I was relieved that I was still alive and, perhaps, because I was a bit wiser. I had had enough of IEDs, SAF prickling at my men, and the occasional lobbing of Taliban mortar rounds into my FOB. It was time for me to bring the fight *to* the enemy. But, to do so, I needed to use a mixture of audacity, guile, and creativity. I also needed some solid military intelligence. And, I needed a different perspective on the enemy.

Fortunately, on a separate portion of my FOB, was an A-Team of US Army Special Forces - the Green Berets. Now, being a conventional Army soldier, I had had almost zero interaction with these fine, unconventional warriors. Admittedly, I was always a bit taken-aback by their swarthy look, their faces bearded and arms sleeved in tattoos. But, I also knew they had a different perspective on the enemy because their primary mission was training their own contingent of Afghan Commandos. I rarely heard about their missions. I think most of them were conducted in other districts and in the dead of night.

I clanged on their gate and the "soldier" ushered me in. (I use quotation marks because his attire was a mix of uniform, local garb, and civvies and he looked like a hitch-hiker from Washington State.) I asked to see his commander and he escorted me to the team sergeant, Master Sergeant Cody Jarrett. I found him in the gym pumping iron to the thrash-metal song "Bulletproof" by the band 5 Finger Death Punch.

He looked at me askance when I asked him for help, but ushered me into his planning room.

"Sergeant, I want to take the fight *to* the enemy; what would you recommend?"

He pondered my question, rubbing his beard and spitting Copenhagen into a filthy cup.

"Attack the enemy's center of gravity."

Now I was familiar with the concept of center of gravity after reading Clausewitz' *On War*. Still, I wondered how could an enemy who is all around us and blended in with civilians have a 'center of gravity'?

Feeling dumb, I asked, "And where is the Taliban's center of gravity?"

"Well, with these bastards, it tends to move from time to time. Right now, the village primarily used by Taliban and their associated terrorist brethren - Lashkar-e-Tayyiba, al Qaida, etc. - is in the village of Tsangar. That's where the Taliban shadow-governor, Nur Akbar, resides."

"Tsangar?!" I exclaimed. "Why, that's only 3.5 kilometers from FOB Honiker-Miracle!"

"Yup. That's why you get pounded by mortars constantly."

"And, that's also uphill in the mountains, at least 2000 feet above us."

"Yup, that's why they keep shooting down at you."

"And, how do you know this anyway?"

The crusty old sergeant leaned forward in his chair and ushered one word, "Spies."

So there it was, he had his own spy network; his own intelligence sources. I was reluctant to delve any deeper; his world seemed just too spooky for me to enter. But, I took his word at face value.

"So, again, how do I get up to Tsangar?"

"Well, hell, Cap'n - y'all are light infantry, aren't you?"

"Well, yes." I stammered.

"And you wear camouflage, don't you?"

"Well, yes, we do wear the latest digi-cams."

"Then do what infantry are supposed to do - hump it. Sure as hell can't get a MRAP up to Tsangar."

What he said made sense, but sounded dangerous. I wanted to seem tough, so I had to use euphemisms to phrase my question properly.

"Wouldn't a dismounted assault lead to a higher risk analysis rating."

He signed. I think he read my mind before I opened my mouth.

"First, the MRAP is a big target. You are a smaller target. One score for lower risk. Second, they've never seen a American dismounted assault. This will shock the crap out of them. Second score for lower risk. The score is now Americans 2, Taliban 0. Now, man-up and go get those bastards!"

I left deep in thought. There was much simple wisdom in what he said. The MRAPs were a blatant target. The enemy would expect us to use our fancy "tanks." Our new cammo patterns really did blend in well. What the heck good were the best cammis in the world if not to *act* as camouflage? Otherwise, the digi-cams were just ugly pajamas. A plan formed in my head.

That afternoon, I called for a shura through the local (corrupt) district governor. I dictated to him the time, noon - to hell with their prayers - but not yet a place. He bristled, but eventually relented. He knew his elders were hungry for free benefits from "Uncle Sugar." The day of the shura, I told them were I wanted to link up. It would be outside the FOB in a field next to a lone cedar tree and a small trickling stream. It was also well within sight of our snipers, who were posted in the guard towers of our HESCO[1] fortified FOB. It was like a pastoral scene from a

---

[1] HESCO baskets are semi-rigid, rapidly deployable, earth-filled defensive barriers made from wire and canvass.

Monet painting. And, it was a safe, unexpected place to conduct an outdoor meeting. (No way in hell was I going to let them peer inside my FOB and see where my mortars and ASP were located.)

As always, the shura was a glorified bitch-session. The old men, their eyes cloudy with cataracts, were gracious at first, but they just wanted free stuff from American taxpayers. It was like they were classically conditioned to accept handouts. I didn't give them much - except a beautiful lie. I told them we were going to patrol the Matin village bridge in two days. This I figured would give their Taliban sons enough time to plan an ambush there and reduce the number of men they had in Tsangar. It also provided me time to devise my assault.

Like clockwork, the morning of the second day, we received radio reports from a patrol out of FOB Blessing, that they received a large volume of SAF as they approached Matin from the west. Now was it my turn. We would raid Tsangar.

My two best platoons had already slipped out of the wire. Intelligence indicated most of the Taliban fighters woke at sunrise to pray, then returned to sleep after a quick breakfast of goat milk and flat bread. The women and children did all the work in the fields; like male lions, the Taliban fighters were unemployed and lounged around when not actually attacking. We surmised they'd be resting their pakol laden heads again at about 0630, two hours after the morning prayer and at that time, we began our ascent. I was with them in the middle of the dismounted formation.

We used satellite imagery to select the route before we launched our ascent. We stayed off the main goat trails, lest a herder compromise our position. We used old-school hand and arm signals to make as little noise as possible. I had informed higher of my plan in advance (I was shocked they approved it!) and had two Apaches on station at nearby FOB Wright to aid in our assault. Also ready were two Blackhawk medic helos, ready to aid the extraction of any of my men lest they become wounded.

Though we moved quietly, our ascent was far harder than I had anticipated! I had expected we would be on the outskirts of Tsangar in an hour. Yet, I had grossly miscalculated the time-distance factor of humping tactically uphill. By 0800, we were still still more than a kilometer away! My men were panting heavily. My legs burned like they were on fire. We were in great shape, but these mountains were horrible! We were constantly sliding and tripping on rocks. Our body armor and ammunition weighed us down like a 100 lb. millstone on our shoulders. Light infantry - what an oxymoron!

We got in assault positions around the village a little after 1000 hrs. I divided our force into three elements - two assault elements and one support element of machine-gunners and mortarmen. Our ROE (Rules of Engagement) stipulated we were not allowed to enter any of the village buildings without provocation, but anyone with guns were fair game. Tsangar sat amid a thick forest - the locals called it "jungle" in Pashto - great for cover and concealment.

And then I saw them - plain-clothed military aged males milling about. We were late and they were no longer asleep. Some of them had AK-47's dangling from their shoulders; others did not. Were they civilians or terrorists? I was amazed at the construction of their hovels. On the imagery everything looked so flat. But here up-close, I could see the structures were built vertically into the sides of the hills! A tiny but prominent mosque was in the center of the village, designated by a loud-speaker on a pole. I made sure to reinforce to my men that the mosque was strictly off limits.

Suddenly, one of the "civilians" must have seen us. The report of a single AK-47 round echoed across the valley. I broke radio silence and ordered the assault. The sweet sound of multiple M4 rifles popped in through the air. For the first time in our deployment, it was *we* who were attacking! My men were pumped. We had not caught the enemy completely off guard, but we still achieved the basic elements of surprise. Armed insurgents fell with shock in their eyes. Slowly, we squeezed the vice using interlocking fires from the two assault elements. I really had little control over the situation. The forest made it too hard to see who was where and what was what. My radio crackled with so much chatter, I couldn't get a clear picture of what was happening. Yet, I knew squads were bounding uphill fighting both gravity and AK-47 rounds. Our SAW machine guns cut down running "squirters" like chainsaws.

Despite the initial success, our assault slowed down rapidly. My men were stoked from caffeine, nicotine, and adrenaline, but kept falling on the rocks, twisting and breaking ankles under the crushing weight of their body armor and extra ammunition. The first few buildings on the outskirts of the village were slowly surrounded and secured, but now the insurgents were fighting back in earnest. Unencumbered by body armor, they seemed to move with speed of a jack-rabbits and the agility of billy-goats. Large 7.62 mm rounds crashed around us, splitting bark from trees. Several of my men tumbled downhill. Hit by rounds and knocking them off balance, they became stuck on their backs like upturned turtles. I heard the deep staccato of an enemy PKM machine gun. It seemed to be coming from the middle of the village, but we weren't certain. Now our assault was at a standstill. Most of my men were pinned behind rocks and massive trees. The enemy was using a technique called plunging fire - firing downhill (often from homes) into my men from high-ground above. CAS (Close Air Support) from A-10 Warthogs was denied because the ROE wouldn't allowed buildings to be bombed. I called in the Apaches who were quicker to react and arrived in five minutes. My men called in the locations where we thought we saw the enemy. The Apache chain guns sprayed lead like a fire-breathing dragon. One squad sergeant finally discovered the PKM location. It was firing from the top window of the mosque! Bastards! They were using the cover of their religious center like a bunker. Then a volley of RPGs snorted from the mosque! We asked the Apaches to fire on the mosque, but they dithered. They replied they needed approval from *their* higher. The three minutes it took to get approval to fire on the mosque seemed like an eternity. My men were going down in scores, pinned by an unseen enemy. Finally,

I saw a volley of Apache Hellfire missiles fly through the air and penetrate dead-center of the mosque's upper room. After that, the fighting died down. An eerie silence blanketed the village.

Exhausted, we were never able to completely surround the village. The earth's gravitational pull worked against us. We tactically questioned the old men we found after the battle. Women wailed and cried. Young boys made us tea. "Where is Nur Akbar," we asked, "Where is Nur Akbar?" Our interpreters, listening in on the Taliban ICOM[2] radio chatter, knew he had been present. They overheard him barking orders during the battle. But, he was here no longer. He had vanished up-hill, likely deep into mountain caves 1000 feet up along with the bulk of his forces. The terrorist "center of gravity" had moved.

---

[2] ICOM is the brand name for a Japanese-made, hand held, two-way radio, popularized by the Taliban.

Despite the protests of the old men with hate in their eyes, I demanded to enter and inspect the mosque. The interior was plain, with rugs and pillows scattered about. In the corner was a huge pile of blue blankets stenciled "UNHCR."

"Will you get a load of this crap?" a young PFC exclaimed at the pile of blankets. He shuffled them aside and there was the motherload - a pile of 20 AK-47s, about

400 rounds, 10 loaded RPG, and three suicide vests!

We tallied the score. Sure, generals constantly pounded in our heads that "we don't take body counts - it doesn't matter" but that was bullshit for us on the ground. The score did matter. What's the point of playing the game if the score doesn't matter? In all, we found nine dead enemy bodies. We discovered blood stains indicating additional wounded insurgents had been dragged away. We arrested another ten suspected bad-guys. Our count? Out of sixty assaulters, two men were killed in action, a third died of wounds, six had other wounds to extremities, two broke their ankles, one broke his neck falling down, eight men were out from heat exhaustion, and ten sustained Class III sprained ankles. It really felt good going on the offensive for a change, but I couldn't call it a real victory. Inkblots Theory[3] be damned, there was no way we could hold the terrain we had just won. At best, it was a draw.

The wounded were MEDEVAC'd by Blackhawk helicopters and those of us who were still ambulatory began a slow trek down the mountain. Walking downhill, I glanced over my shoulder at the enemy's former

---

[3]Inkblots COIN Strategy - The Inkblots counter-insurgency (COIN) strategy was successfully employed by the British during the communist Malaysian insurrection from 1948-1960 and largely developed by Sir Robert G.K. Thompson. In Afghanistan, it was popularly known as "Clear-Hold-Secure" which describes how it works. Once you clear, hold, and secure a large section of geography, this secure area should expand into other secure areas, like inkblots slowly covering a page. In contemporary times, the theory's main proponent has been LTC(R) Andrew Krepinevich, a Vietnam veteran.

strongpoint, the village mosque. I should have focused downhill, for I stubbed my toe on a large rock and came crashing down, my jaw catching a boulder. Stars danced around my eyes and everything went black.

<u>Lessons Learned:</u>

- *Its easier to fight going down a mountain than uphill - seize the high-ground before the enemy.*
- *Surprise equals success. You will never surprise the enemy if you keep doing the same damn thing. Don't be afraid to mix it up, even if 'fraidy-cat colonels consider it unorthodox.*
- *Light infantry needs to be l i g h t. Light = fast; fast = mobile; mobile = lethal; lethal = they die, you live.*
- *Satellite imagery is a nice tool, but its not the "be all and end all." You must be on the terrain to understand the terrain. Don't confuse modern tools with superiority.*
- *Whether you call them insurgents, enemy combatants, terrorists, or bastards, they're still all* civilians *- not an army; not internationally recognized soldiers.*
- *Let me be clear - <u>all terrorists are civilians,</u> even though <u>not all civilians are terrorists.</u>*
- *ROE's are made by lawyers to prevent generals from going to jail. They are handcuffs on you and don't win wars.*
- *Remember this and take it to heart – you may train hard at JRTC, NTC, CMTC, etc. but your enemy trains on <u>real</u>*

battlefields. Only the fittest survive his training and his next training session is against you!

- *In Islamic countries "Mosque" has five definitions: a) place of worship b) place where terrorists hold meetings c) propaganda center d) weapons depot e) sniper position.*

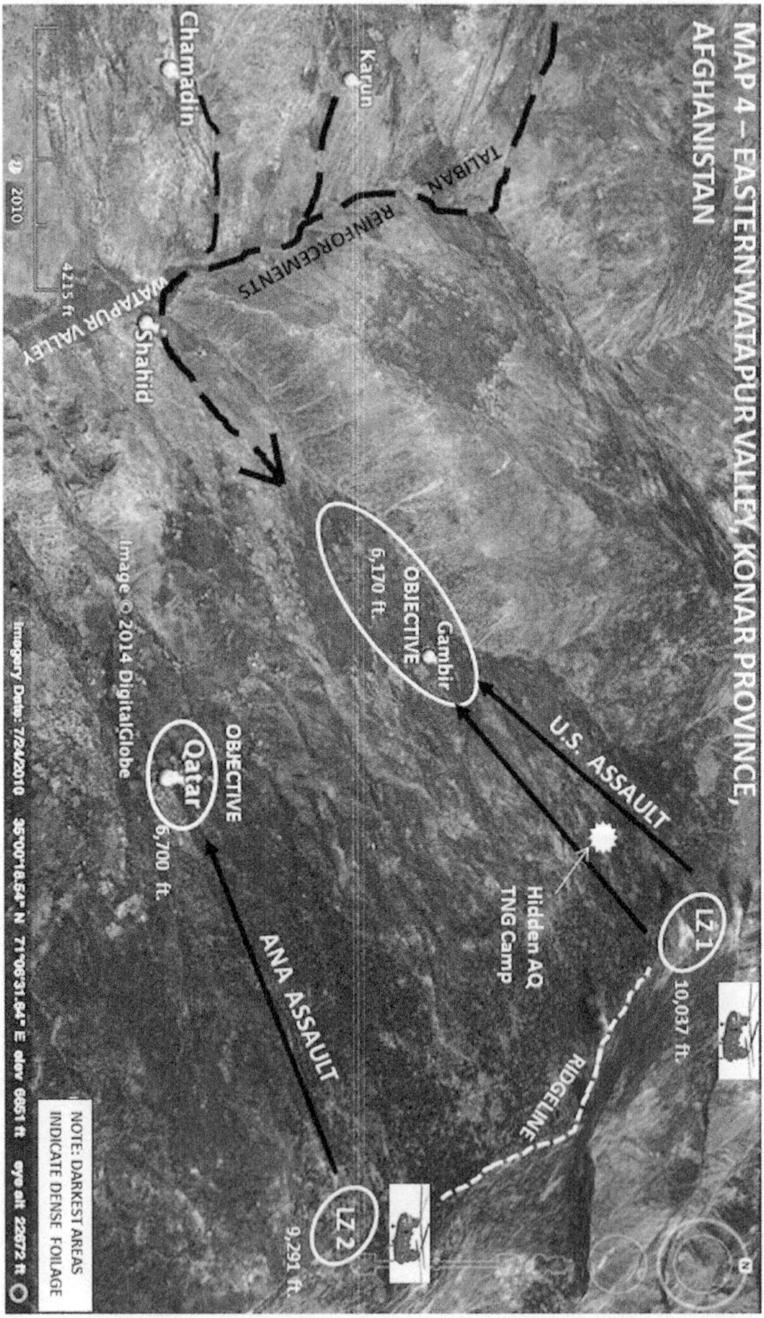
MAP 4 – EASTERN WATAPUR VALLEY, KONAR PROVINCE, AFGHANISTAN
TALIBAN
REINFORCEMENTS
Karun
Chamadin
WATAPUR VALLEY
Shahid
4215 ft
2010
Gambir
OBJECTIVE
6,170 ft.
Image © 2014 DigitalGlobe
U.S. ASSAULT
Hidden AQ
TNG Camp
LZ 1
10,037 ft.
RIDGELINE
OBJECTIVE
Qatar
6,700 ft.
ANA ASSAULT
LZ 2
9,291 ft.
NOTE: DARKEST AREAS
INDICATE DENSE FOILAGE
Imagery Date: 7/24/2010 35°00'18.64" N 71°06'31.84" E elev 6651 ft eye alt 22872 ft

CHAPTER 4 - Fresh Ideas

I awoke from my slumber feeling refreshed. I had had deep dreams of battling my way uphill in heavy body armor and getting shot at. I was glad it had all been a dream, the deep slumber courtesy of Ambien. (I had long ago quit my Mefloquine anti-malarial pills - they really gave me bad dreams.)

Six months into my year-long tour, I had quickly grown sick of watching my men mortared, blown up by IEDs, being drawn into ambushes by B-S KLEs, and sniper fire. It was time to stop playing defense. The enemy seemed to anticipate our every move, and knew our ROE better than we did. It was time to go on the offense. The question was, "how?"

For fresh ideas, I went to visit with the Special Forces A-Team on the other side of the camp. I made friendly relations with these unconventional warriors a priority when I first arrived, but I had never asked them about tactics before. I found the team sergeant, MSG Cody Jarrett, in his day room pouring over intel reports with the Disturbed's song "Indestructible" blaring in the background. I explained to him my desire to go on the offense.

"So, MSG Jarrett, how would you go about it, if you were in my shoes?"

MSG Jarrett stroked his swarthy beard and took another pinch of the local gooey, green Afghan tobacco, *naswar.* "Can you get CH-47 Chinook support?"

"I think so." Actually, I had never asked higher for helo support before.

"Then take the high ground. Work your way down the mountain."

Of course! Air mobile ops! Elementary, Dear Watson! Jarrett continued his questioning.

"Y'all got nods?"

"Yes, we have some good night-vision equipment."

"Have your men trained in them?"

"Why, yes." Truth told, we had trained in nods, but not since our pre-deployment training almost a year ago at JRTC. My men were rusty at using them, but capable. I made a mental note to start using them here, in theater, and get my men comfortable with them ASAP.

"Good, Cap'n. Then use them. They own the mountains, but we own the night."

"But, where...where should I attack the enemy to best cripple him?"

"Well, Cap'n, my latest HUMINT (human intelligence) reports tell me Nur Akbar and his al Qaida trainers are holed up in the village of Gambir."

"Gambir?" I exclaimed. "I thought Nur Akbar was from Tsangar."

"He has three wives. He keeps one wife at his home in Tsangar, one in Gambir, a third deep in the mountains northwest of the Watapur in Qaylaygal. Right now, he's in Gambir."

"Really? Okay, but I gotta ask, how reliable are your HUMINT reports?"

"One of my spies is Nur Akbar's first cousin. He wants Nur Akbar dead because Nur Akbar is playing *bachabazi*[4] with my spy's teenage son. Plus, my 18F (SF intel MOS) pays the spy very, very well."

"Ewww. Wow - I'd want him dead if he were buggering my son, too."

Fresh with this new information, I set the wheels in motion for my assault on Gambir. The first thing I did was order my squads to start conducting small unit training at night on the FOB, to include night fire. Simultaneously, I sold my plan to my commander and convinced the Army bureaucracy to start planning CH-47 support for my operation. Both would take about a month to prepare. I wanted the op to take place when there was little or no moonlight so we would have the advantage against the enemy at night. The aviators had the idea to bulk up night logistic flights to the FOB for a month, so it would appear as if night flights were a new normal pattern. I also ensured everyone boots were inspected. Boots with poor ankle support would have to be replaced and broken in. Lastly, my NCO's inspected everyone's gear to the smallest detail. We needed to be as absolutely light and nimble as possible. Every bullet would have to count. Every ounce off their gear that wasn't absolutely necessary was stripped away. I ensured the mortars and artillery were properly registered. So, every time we received a n enemy mortar

---

[4] *Bachabazi - the Pashtun practice of pederasty. Yes, that's right, not only are you fighting Islamic fanatics, you're fighting child molesters, and drug addicts.*

round, we would lob rounds against the mountain to register the big guns for the Gambir area, not necessarily try and destroy random enemy mortar positions. Finally, a week before the mission, higher convinced a regular infantry company from the Afghan National Army to participate in the operation. This would put an "Afghan face" on our mission.

Gambir would be a tricky village to attack. It was actually a cluster of three hamlets - upper, lower, and middle Gambir. This part of the Watapur valley was also extremely forested, like the jungles of the Pacific Northwest. Our plan was to come in by multiple Chinook lifts and drop off our infantrymen right on the very top of the ridge-line that dominated the eastern side of the Watapur Valley. In addition to regular lifts, we would implement several "false" insertion points, as a ruse to confuse the enemy. We would assault down the mountain towards Gambir; the ANA would tackle the easier mission and sweep down towards Tsangar. Lastly, I wanted to be sure our bold mission was recorded. So, I invited a respected American reporter to be imbedded with our men during the assault.

The day of the mission, I had a chance to inspect the ANA soldiers. I was so proud to have them in on this mission. Their faces were the multi-cultural faces of the future of Afghanistan: a mixture of Pashtuns from Helmund Province in the south, Hazaras from central

Afghanistan, and Turkomens from the north - all united with one mission to tackle a terrible foe!

Midnight arrived - H hour - and we were off into the black of night! After one false insertion, my bird landed us safely on the crest of the mountain. Though the middle of summer, there was a chill at this altitude of 10,000 feet, something I didn't expect. Slowly hiking down the mountain, I also noticed how thin the air was. This made breathing harder than I anticipated, but at least we were going downhill! Our interpreters heard the Taliban talk up a storm on their ICOM radios. It was readily clear we had total surprise. A litany of Taliban group leaders, awoken from their slumber kept exclaiming,

"Oh my God, the Americans are here. Hey, can you see them, Ali?"

"Holy crap, Mohammed, no! They seem to be everywhere?"

"What shall we do, what shall we do?"

"Are they building anything on top of the mountain - do you see HESCO baskets?"

"No, no, shit, I can't see anything!"

The enemy was completely confused.

It was very, very difficult to navigate our way in the dark, in the forest, and keep all of our elements moving in a coordinated, organized fashion. But, by first light, we had upper Gambir mostly surrounded. My orders were clear: the first man you see with a gun - shoot and kill him. And, sure enough, by first light, they were there. The enemy was ready for a fight - and so were we. Chaos ensued. Bullets cracked through the air. We tightened the noose, doing 3-5 second rushes from rock to rock and tree to tree. Mortars landed with deadly accuracy, eliminating small pockets of Taliban. We moved forward and fought house to house. It was exhausting but exhilarating! The Taliban were in the fight of their life. All be damned if every military aged male didn't have a rifle. From age 14 to 60, down they dropped, pakols and man-dresses splattering red. Some, after being shot, jumped back up like zombies and charged. They were likely on opiates. With head-shots, we but down those zombies for good. Near the center of the village, I saw a woman in a burkha pick up an RPG and start firing over her dead husband. Another woman emerged from the house firing on my men. My men were reluctant to return fire, but a brave NCO with multiple Iraqi tours barked out order to wax the bitches and down they went.

The battle continued into the following day. With Gambir now secure, my men reported hordes of Taliban reinforcements hiking up-hill to repel us. Intel reports stated they were from far-away Qaylaygal and Mandesh villages, but I found it difficult to believe their reinforcements could move so far so fast. In waves they came, prime targets for our M240 machine guns. Suddenly, we had contact to our rear, from the forest uphill! How could this happen? Fortunately, I had third platoon in reserve. They fixed these fighters, then called in scores of 155mm fire missions. Damn, it was so hard to see them in the forest. Third platoon maneuvered uphill to where the resistance was coming from and decimated the insurgents. Combing the pockets of the dead for documents, we made an interesting discovery - passports from Pakistan, Saudi Arabia, Egypt and Russian Chechnya. We also found lots of syringes; proof many of these "holy" fighters were doped-up on drugs. Later, we found the nest's epicenter. They had caves shrouded with tents of green. Our satellite imagery had not detected this under the heavy foliage. It was clear what had been behind us, uphill from Gambir. In the dense foliage, we had walked right by an al Qaida training camp.

In all, we fought on that mountainside for six long days. Dead Taliban bodies lay rotting in the sun, for our fire was too intense for the enemy to recover their dead. Our terps heard their ICOM radio chatter bemoan their rotting dead. *"Wakil, if our dead don't get an Islamic burial in 24 hours their souls may go to Hell!"* Pity. Finally, our commander called us home. We couldn't secure the area. That job was now left to the ANA, who had encountered little resistance in Tsangar.

Debriefing my men afterward at the FOB, we tallied the score. In six days, we killed 125 insurgents, including seven Arabs, two Pakistanis, and one Chechen. Intel reports indicated Nur Akbar was wounded on Day One of the battle, but somehow had slipped away. Our casualties: one man died of wounds after breaking his neck in a hard landing inside one of the Chinooks, four minor flesh wounds, three Class I sprained ankles. Our commander called it the most decisive victory against the Taliban every accomplished in Eastern Afghanistan. Then the manure hit the fan.

LTC Lemoux summoned me to Jalalabad. He was furious.

"Have you seen the news!? Have you seen the news!?" he shouted at me.

I felt like telling him, "*No, LTC REMF*[5] *- I live out in the boonies. I don't get CNN out there*," but I held my tongue and gave him a curt, "No, sir."

Our friendly imbedded reporter wrote about the battle. The headline read, "Americans kill numerous women in the village of Gambir." The reporter went on to state he had been an eye-witness of the "atrocity." It was the talk of the town back in the USA. No where in the article did the reporter mention the women were firing at my men. One prominent senator with lofty presidential aspirations waved the newspaper on the floor of Congress and chided the Army for "killing women and children in the dead of night." ALL offensive actions were canceled pending a formal investigation. The soldiers who killed the women were sent Stateside in handcuffs to testify. Luckily, one of the men wore a Go-Pro helmet camera and recorded the incident in full. They were all exonerated after six months in the brig, but the damage done. But, things were worse than that.

---

[5]*REMF - A profane acronym that stands for Rear Echelon Mother F@#ker*

Five weeks after our victory, I got a surprise visit from the ANA commander.

"We're leaving - we're done."

"What? What the hell do you mean 'you're done'?"

"Listen, I am only at 50% strength now. My ethnic Pashtun soldiers have all deserted to the hills. My ethnic Turkomen began harassing the ethnic Hazarras and cut off their food supply. The Hazarras began trying to beat up the Turkomen, who responded by shooting two of the Hazarras. I have no 'force' now. I'm taking what's left of my company and headed to Kabul."

A week later, we received a massive artillery barrage. Engineers, analyzing the angle and trajectory of the rounds determined they were from enemy mortar positions in Gambir. Over ICOM radio chatter we heard the ugly voice of Nur Akbar, praising his men for bravely retaking Gambir and cursing the cowardly Americans infidels. His speech was delivered on Friday morning, during a religious sermon at the local mosque.

I squeezed my eyes with frustration. Oh, how I longed for a good stiff sip of alcohol! But, I had none, courtesy of "General Order-Number-One." (I'm sure General Order-Number-One was probably drinking martinis on a golf course near the Pentagon.) Exhausted, but too frustrated to fall sleep, I took a double dose of Ambien that night, hoping my dreams would smother reality.

<u>Lessons Learned:</u>

- *One well placed spy is worth a thousand warriors.*
- *If you don't <u>practice</u> owning the night, you won't.*
- *Head-shots kill zombies.*
- *Watch out women - if you're sleeping with the enemy, you're fair game.*

- *The press is not your friend. If you can't manipulate them to your own purposes, get them off the damn battlefield.*
- *Americans are the masters of media and media should be used as a positive propaganda tool. Propaganda is an*

essential element of warfare. If you don't use it, the enemy will.

- *There can never be truly "national forces" in a tribal land where there is no sense of nationalism. National forces in tribal society won't attack Islamic terrorists that don't effect their family or tribe. Only loyal local forces give a damn.*
- *Hey Mister Politician - I've got a great British phrase for you - PISS-OFF!*

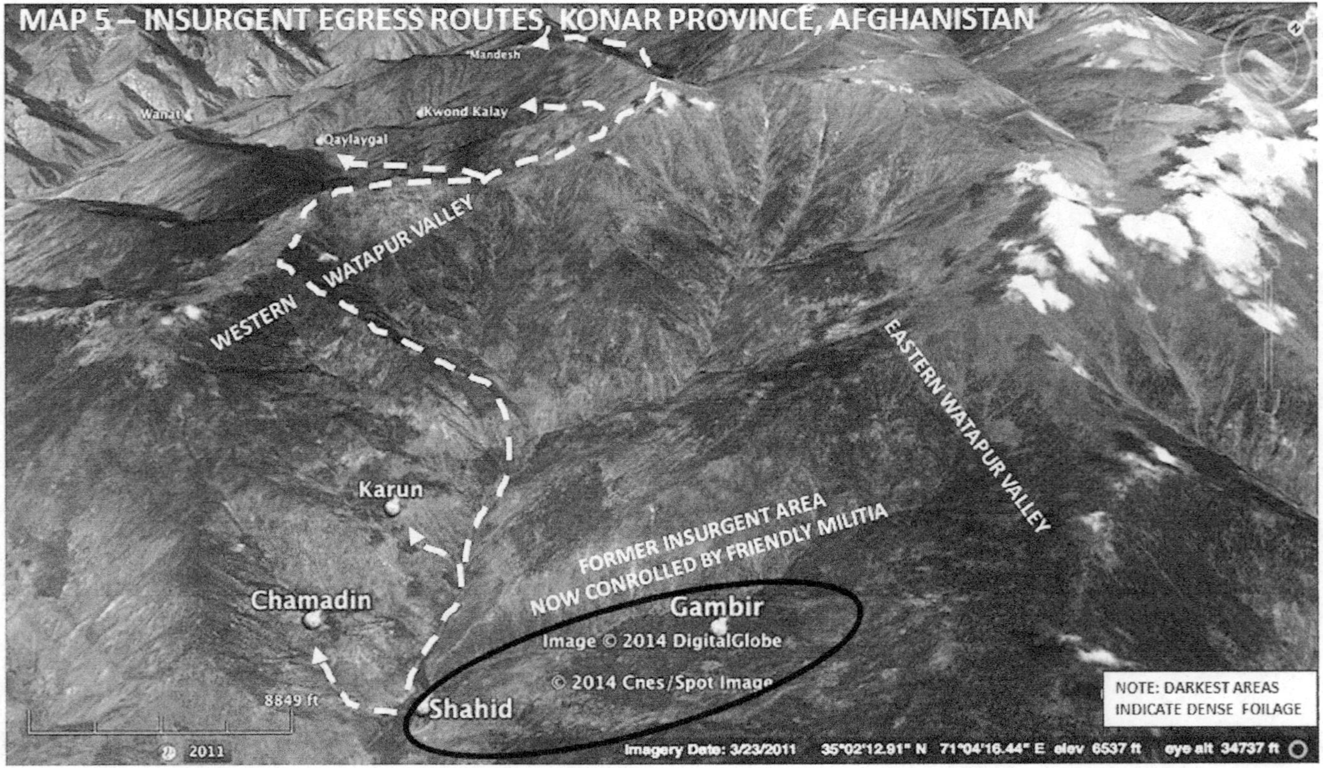
MAP 5 – INSURGENT EGRESS ROUTES, KONAR PROVINCE, AFGHANISTAN
Mandesh
Kwond Kalay
Wanat
Qaylaygal
WESTERN
WATAPUR VALLEY
EASTERN WATAPUR VALLEY
Karun
Chamadin
FORMER INSURGENT AREA
NOW CONROLLED BY FRIENDLY MILITIA
Gambir
Image © 2014 DigitalGlobe
© 2014 Cnes/Spot Image
Shahid
8849 ft
2011
NOTE: DARKEST AREAS
INDICATE DENSE FOILAGE
Imagery Date: 3/23/2011 35°02'12.91" N 71°04'16.44" E elev 6537 ft eye alt 34737 ft

CHAPTER 5 - Spies

I woke with the grit of sand in my eyes. Wow - what a deep Ambien-induced dream I had just had...or *was* it a dream? I dressed quickly and went to the TOC. I asked the NCOIC[6] if there had been any changes overnight. Nope. We still didn't own the valley. I asked him about our airmobile op in Gambir, but he looked at me as if a horn had spouted from my head. "I don't know what you're talking about, sir, but it sure would be a great idea," he replied.

Perplexed at this dream inspired idea, my mind searched for the right chess moves to make this happen. What would be the best way to go on the offensive and secure terrain with limited resources? For answers, I marched over to my Green Beret friend, MSG Cody.

I found Cody eating local yogurt with a bearded man in a pakol and some wicked song from the metal band Avenged Sevenfold blaring in the background. Cody introduced the man as "Musa." Musa greeted me in the typical Konar fashion by putting his hand over my heart, and I returned the local greeting.

MSG Cody got right down to business. "Cap'n, I'm glad you came by. I was just chatting with Musa some plans we've developed together and we could use your help," he remarked.

---

6 NCOIC - Non-Commissioned Officer In Charge

"Good, because I have an idea, too, I wanted to share with you. But, first, could you please explain what Musa is doing here and who he is?"

"Musa, Cap'n, is the head of the *Arbakai* force we've been training for the past several months."

"Arbakai?"

"Yes. Local militiamen. As you probably know, the traditional mission of the Green Berets is to train local, indigenous forces. So, for a long time now, we've been training both our own Afghan special forces, whom we call Commandos, *and* local militias, the Arbakai."

"Ah, yes, of course." I replied, faking my full knowledge of his mission and tactics. "I had read about such tactics used in Vietnam."

"Well, Vietnam is not the only time we've done this. The Brits did it for years during their colonial wars. We Americans also used similar tactics in Nicaragua during the 80's, the French in WWII, and as far back as the American Indian Wars with tribes like the Crows and Tonkawas."

"So, why do you think the Arbakai will help?"

MSG Cody smirked at my stupid question.

"Cap'n, you ever seen the 1984 movie *Red Dawn*."

"Uh, sure, that's a cult classic for military men." I said this realizing Cody had probably seen it as a teenager on the big screen, while I saw it courtesy of Netflix.

"You know why those teenagers in the movie fought so hard against the Soviets and Cubans?" Cody asked.

Probing the deep recesses of my cerebral cortex for the answer, I recalled a scene where the father of a teenage insurgent screamed, "*Avenge me son!*"

I answered meekly, "For revenge?"

"Partly, right, Cap'n. It was for revenge. But, they fought hard and were successful because they were fighting for *their* local community. You can't get a Tajik from Mazar-i-Sharif to fight for a Pashtun in Konar. This country doesn't understand the term 'Nation-State.' Few in the third world have such a concept. You have to tackle this *one tribe at a time*[7]."

Together, over strong Afghan tea and naswar, MSG Cody, Musa, and I worked out a plan. We would initiate an Airmobile operation, taking the high-ground over the eastern valley of the Watapur valley. We would be the spearhead. MSG Cody's Commandos would then fall in behind, and clean up the villages of Tsangar and Gambir proceeding building by building, cave by cave. Following the Commandos, the local Arbakai would be instituted as the new community leaders. We would depend on them to secure the populations...by *any* means *they* deemed necessary according to *their* customs.

After rehearsing for several weeks, we launched our assault in the middle of the night, seizing the high-ground of the Eastern Watapur Valley and working our way

---

[7] This phrase is taken from the 2009 book, *One Tribe at a Time: A Strategy for Success in Afghanistan*, by Major Jim Gant, a book I highly recommend. (See my recommended reading list in Part II.)

downhill. It was bloody, difficult work. In six days of intense fighting, we killed over 200 insurgents. Near the final day of the battle, another company in our battalion launched a very rapid ground assault up the valley to secure the villages at the bottom of the mountain in an attempt to cut off any retreat by the enemy. The AH-64's did most of the damage there, decimating "squirters" trying to escape from Shahid village across the valley floor. This proved difficult for many unarmed civilians were also fleeing. Those who fled were obviously pro-Taliban; hard-core supporters. The villagers who stayed behind welcomed us and the Commando's as great liberators.

After the battle, Musa's men uncovered tons of Taliban, Al Qaida, and Islamic State propaganda videos.

These were horrific videos of evil men torturing men, women, and children. I did not allow the media to accompany us during the battle, I made certain to give the reporters the propaganda videos after the battle, but only *after* I had *forced* them to watch the beheadings and torture videos. I didn't care if it made them sick, I wanted them to know what US troops were fighting *for* and fighting *against*. I told them, "Put *that* in your newspapers, Mr. Reporter. And while you're at it, make sure you send some copies to Senator Such-And-Such." Some of the reporters were angry with me for forcing them to view these atrocities. But, *some,* (maybe the few honest reporters), maintained an open mind, (or at least believed the videos would sell more papers or draw traffic to their blog sites.)

We did lose several men in the battle. This always made me very emotional. I had to be the one to write condolence letters to the widows and mothers back home. But, this was the price for freedom and the cost of war. Warfare, by definition, is not a zero-defects event.

Overall, the operation was a tremendous success, with two exceptions. Nur Akbar and his Arab AQ operative, Barak al-Qahtani, escaped our noose. They fled to the hills and valleys northwest of Shahid.

For about three months afterwards, relative peace ensued. We continued to conduct presence patrols, but no one shot at us. The Arbakai continued to exchange sporadic fire with Taliban in the hills, but that was to be expected. Then the calm ended.

"Sir, sir, you've got to come to the front gate and see this!" an exasperated NCO rushed into my office and exclaimed.

"What is it, Sergeant Trinidad?"

"Just come now, sir...you gotta see for yourself!"

I went to the front gate and there was a great commotion. A large crowd of locals had gathered and were muttering in inexplicable tongues and pointing and gasping. I saw immediately what the commotion was about. Four fresh heads were lined up in a neat row on the ground. I recognized one of them. It was Musa. A note was pinned to his nose. The terps translated it.

> "HERE LIES THE HEADS OF INFIDEL AFGHAN SPIES SUPPORTING INFIDEL AMERICANS. WE WILL NOT REST UNTIL THE GREEN FLAG OF ALLAH AND HIS PROPHET MOHAMMED, PEACE BE UPON HIM, IS RAISED OVER THE WHITE HOUSE OF WASHINGTON, BUCKINGHAM PALACE, AND TOPS THE EIFFEL TOWER! ALLAH IS THE GREATEST!"

I met with the leader of the crowd of men. He informed me masked men were ravaging all the villages on the west side of the Watapur Valley near Chamadin village where we had no Arbakai presence. Theses masked murderers were going house to house interrogating people in the dead of the night. Anyone they suspected of supporting the Afghan government were executed. Many were tortured to death, roasted alive over burning fires. Even Islamic religious leaders who allowed girls to be go to school were killed. Musa and his men tried to help some of the villages, but he was captured in the raid and beheaded by the Al Qaida Afghan leader, Mohammed Habib. MSG

Cody came rushing down to the gate in shorts and a T-shirt, his beard flowing in the breeze.

"Shit." he muttered angrily. "One of my spies just told me Musa had been captured."

"Are any of these men your spies?" I asked.

"Not a one of them. Totally innocent men."

My head turned slowly to the crusty master sergeant and I asked, "Who would do such things?"

"Evil men."

"How do you confront such evil?"

"With overwhelming force, cunning, and violence."

"What do we do next, Jarrett?"

"Cap'n, we need to cut the head off the snake - Mohammed Habib, Nur Akbar and that damned Arab, Barak al-Qahtani."

I saw an intense fire in his eyes. Musa had been his friend and a very brave man. Now it was personal.

"AAAALAHU-AAAKBAAAR!!"

I saw the flash before I heard the boom. I felt my body sailing through the air. I landed with a dizzy thud. Shaking my head from the ground, I saw pandaemonium before my eyes. Body parts littered the area where the crowd of civilians had once stood. Pieces of flesh littered my tunic. A suicide bomber had infiltrated the crowd and detonated himself killing a dozen villagers. I vomited at the sight. Then I went blank as I passed out in my own spoilage.

Lessons Learned:

- *Warfare is neither calculus nor physics. Every action has a reaction, but it is not necessarily equal nor opposite.*
- *The enemy doesn't care if they kill their own. They rule by fear. It's the law of the savage. Warfare, by its very nature, is savage. YOU must be a BETTER savage.*
- *Inkblots Strategy has merits, if you use the right ink and the right amount of ink! It will always be a very time-consuming, expensive, and manpower intensive endeavor.*
- *Don't waste time foisting Western democratic values on Eastern, Islamic people. They won't get it. Spend your time killing the savages that want to kill you and training the good Muslims to kill the bad Muslims.*

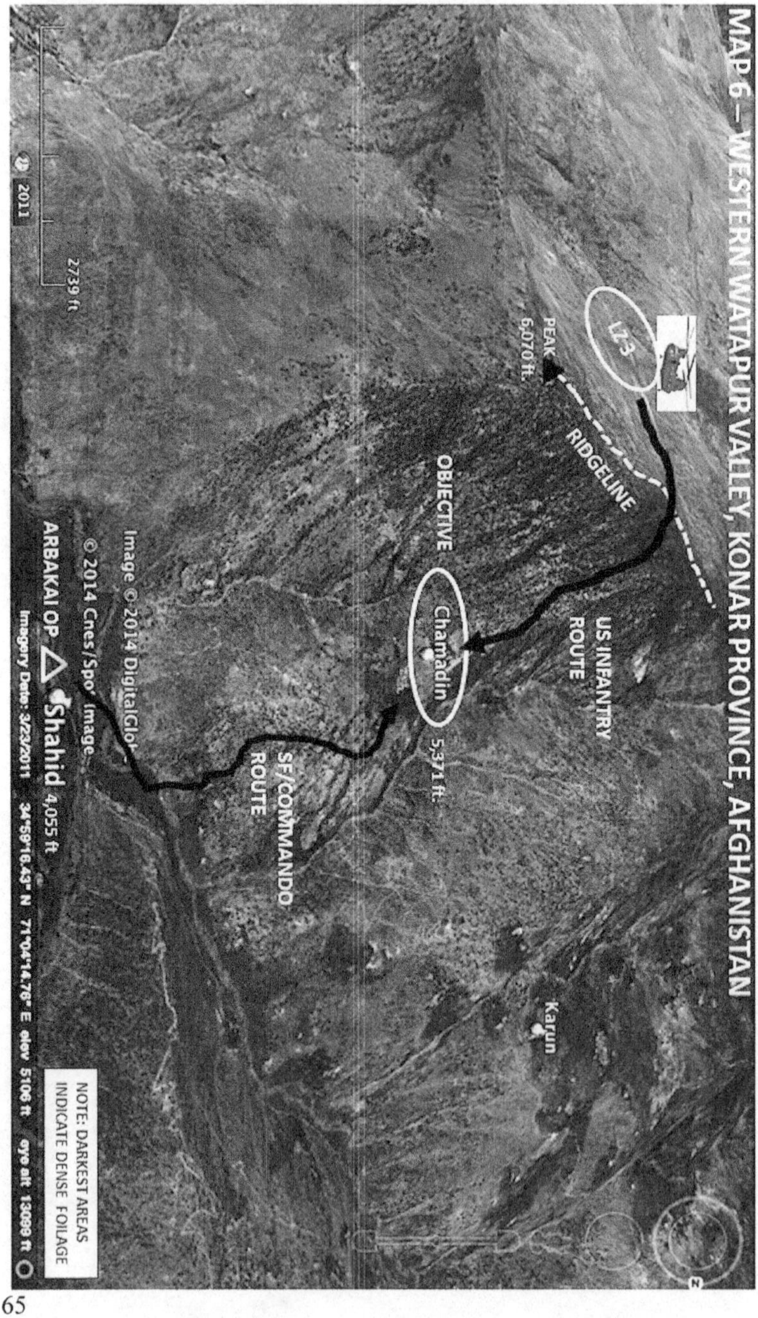
MAP 6 – WESTERN WATAPUR VALLEY, KONAR PROVINCE, AFGHANISTAN
LZ-3
PEAK
6,070 ft.
RIDGELINE
OBJECTIVE
US INFANTRY
ROUTE
Chamadin
5,371 ft.
SF/COMMANDO
ROUTE
Karun
Shahid 4,055 ft
ARBAKAI OP
NOTE: DARKEST AREAS
INDICATE DENSE FOILAGE
2739 ft
2011

CHAPTER 6 - Vipers

Darkness encased me. The smell of old sweat invaded my nostrils. I struggled to writhe from the grasp of irons binding my wrists. I tried to shout, but no sounds emanated from my lungs. Then, a slick coil slowly began to twist around my ankle. A snake! A deadly Afghan viper? With lightening quickness, its fangs pierced deep into my skin and bolted from my bunk with a gasp of panic!

I was still in relative darkness, entombed in the confines of my small wooden room drenched with sweat. I flipped the light switch, my ankle in pain. In throes of my nightmare, I had slammed my ankle against the crude wooden panels next to my crude bed and a nail had pierced my skin. Whew! Not a snake. I looked down at the floor and saw Edgar Allen Poe's book, *The Cask of Amontillado* resting on the floor beside me. "*That's the last time I'll read Poe to try and fall asleep - better to stick to Ambien*," I mused.

The night's dream of blood and bodiless heads still haunted me. Where had I gone wrong? How could I defeat an enemy I could not see? Again, I looked down at the cover of Poe's book, and the coat of arms of Montresor stared back at me. It depicted a shield of azure with a foot holding down a snake. The snake had turned and bit the ankle above it. Below was the family motto: *Nemo me Impune Lacessit* - "no one insults me with impunity." And my mistake suddenly became clear. Using airmobile ops, overwhelming force, and the local militia, I had secured the eastern valley of the Watapur, but my foot was only holding down the serpent. That was not good enough. I had to decapitate the viper!

I dressed quickly and made my way directly to MSG Jarrett's compound, sans body armor and helmet. Finding him listening to Metallica's "Enter Sandman" from their Black Album came as no surprise. My newfound confidence did.

"Jarrett, now I know what you meant about cutting the head off the snake. Right now, we're merely holding him down, limiting his exposure, and chasing them to different holes."

"Ah, yes, Cap'n! The Taliban, Al Qaida, and ISIS are slippery serpents, indeed!" He recognized the new resiliency that must have gleamed in my eyes.

"Is your network of spies still viable?"

"Yes, of course. The insurgents have lashed out at innocent civilians because my men are so well imbedded in their bloodstream."

"I want to go snake hunting, Jarrett."

"Me, too, Cap'n!" he said, pausing to inspect me more closely. "I can see you now have the first necessary element of warfare - you have the WILL. You must have the *will* before you can find the *way*."

"So, what is the way? How do we cut the snake's head off?"

"We must first find out who the head of the snake is, and then find out where he is. We already know some of the *who's* - for this snake is a multi-headed hydra - men like Nur Akbar and Mohammed Habib are only some of the heads."

"So now we need to know the *where*."

Jarrett raised an eyebrow and looked at me askance. "Doesn't your S2 shop ever read the intel reports my 18F's write?"

Now it was time for me to look at him funny. "Are you suggesting you've written intel reports on the 'where'?"

“Ha, ha, ha - yes, many. We know through our network precisely where Mohammed Habib’s home is. This was Musa’s target, before he was tortured and killed. We also know from pattern analysis that Nur Akbar usually meets Mohammed Habib at his home every Sunday.”

“Well, what are we waiting for?”

“Well, now that we have the *will*, the *who*, and the *where*, and the probable *when*, we need two more things - a *trigger*, and a *plan*.”

“A trigger?” I asked. “Explain.”

“We need some means to confirm the snakes are in their lair. And I have just the man - my man Rasaq.”

MSG Cody Jarrett revealed that he had arranged a series of signals with Rasaq. Whenever Nur Akbar was at Mohammed Habib’s home, he would burn a large pile of garbage in a specified location that could be seen by the Arbakai and relayed to his TOC. Smoke signals - right out of American Indian tactics; brilliant in its simplicity and not requiring fresh cell phone batteries or expensive satellites.

I glanced at my watch. Sunday was in 36 hours. There was no time for a long OPORD process or time to staff my plans with my battalion commander or go through long, drawn out company rehearsals. I needed a simple plan that could be enacted fast. So, together with Jarrett, we devised a very simple plan. In fact, it was downright audacious in its simplicity. His Afghan Commandos would cross the wheat field in the bottom of the Watapur Valley from an Arbakai Observation Post (OP) near Shahid village at night and march uphill, west towards Chamadin village, and Mohammed Habib's compound. I would take my most crack platoon and best snipers, conduct a false insertion above Chamadin village and march downhill from west to east. We would surround the compound at night and get the snake napping!

At dusk on Saturday night, the Arbakais reported Rasaq burning his garbage. The "trigger" had been pulled. We lifted into the night and our Chinook circled south, initially away from the target. The commandos, being Afghans, would have no problem marching uphill. We, being Americans, would have no problem marching downhill. We had off-set from the target building by 4KM, so, given our slow march speed in the darkness, we would be scheduled to reach the compound by 0330, approximately an hour before the morning prayer hour. Guided by our GPSs, we approached the compound and used a series of Infrared (IR) flashes of light to signal our approach to the Commandos (also equipped with NVGs), ensuring we wouldn't shoot each other and maintain radio silence.

Most of the hovels of Chamadin were identical, but Rasaq had informed us of a distinct, lone oak tree in the animal pen of Mohammed Habib's compound and had drawn schematics of it for both our elements. Slowly, I watched see the rectangular roof of Mohammed Habib's compound form in the green of my nods. Unpenned chickens roosted in the animal pen under the unmistakable lone tree. My snipers were in position to provide overwatch and fire on any armed squirters trying to flee. I could see the distinct bulky outline of MSG Jarrett towering over his commandos approaching the compound from the east. Using hand and arm signals, we covered every entrance of the compound. It only had three. We tried the main door. It was unlocked. Slowly, stealthily, we tip-toed into the central corridor. Sleeping bodies were strewn about, weapons resting by their beds. The commandos guarded the entrance to the women and children's sleeping quarters. Then, we simply woke them up with a nudge on the shoulder.

The element of surprise was absolute! We didn't even fire a shot! We lined them up quickly, segregating all the men, and zip-tied their hands as our interpreters spat out a series of quick questions in rapid succession.

"Who are you!? What is your name!? Do you have any hidden weapons!? Where!? Who else is here!? Who is that man!? Who is this man!?"

In the shock of our presence, everyone confessed their identity. Even Mohammed Habib and Nur Akbar, though they tried to use their nome-de-guerre's to confuse us. Their cell phones and ID cards further confirmed their identity. Now I wanted to get off the "X" as soon as possible. I radioed in for the Chinook to the nearby rendezvous point and we lifted away into a peaceful morning dawn!

My men did fist pumps and high fives all around! We didn't just get the heads of the snakes, we captured them! My battalion commander and S2 were elated. This would be a marvelous treasure trove of intelligence once they were back at Bagram in the interrogation facility.

Then, disaster struck. The lawyers came.

The ISAF defense team claimed that these men were civilians. We had no cause to confine them. The Afghan government claimed Mohammed Habib and Nur Akbar were farmers. Even though they slept with loaded AK-47s, it wasn't considered "proof" they were Islamic terrorists. We had a body of intelligence to support their years of terrorist activities, but some staff officers bickered we could not release the intelligence reports as part of the prosecution because it would reveal "sources and methods." The Taliban deputy shadow governor paid a bribe to the Afghan government governor of Watapur to try and secure their release. The proceeding ground to a halt. Meanwhile, interrogators couldn't pry a single word out of the prisoners. They claimed to be completely innocent. There had been weapons next to their beds, sure, but they claimed they were self defense weapons *against* Taliban intruders! Six weeks after they were captured, they were released from bondage.

The news crushed our spirit. No one felt like taking chances anymore. In *The Cask of Amontillado*, Montresor killed Fortunado with devilish guile and cunning, living up to his family motto. But our own system had turned the tables on us *in favor of* the enemy. We had become confined to the catacombs of Afghanistan by the rules of engagement that our American politicians and lawyers had created. Fortune was not ours because Fortunado had been freed.

Incoming mortars thudded again in our compound, shaking the dusty ground like an earthquake. I sighed, donned my body armor, and sulked slowly to the bunker, not caring much at all whether I lived or died. One large Soviet round rocked my bunker, bursting my eardrums, and knocking me again unconscious.

<u>Lessons Learned:</u>

- *Lawyers belong in a courtroom, not on a battlefield.*
- *A released terrorist becomes <u>more</u> dangerous because his capture gives him "street cred." Usually released prisoners get a promotion and acclaim for having "beaten the system."*
- *A captured terrorist is only as good as the information he has or can provide you or is willing to provide you. After you've gotten all you can out of him, he's worthless and should be executed.*
- *Dead men tell no tales. Dead men also don't return to fight you again tomorrow.*

CHAPTER 7 - Collateral Damage

I was awakened from my deep slumber by the shouts and hoots of soldiers outside my door. Rubbing the grit from my eyes, I threw on my trousers, quickly laced up my boots, and dashed outside to see what the commotion was all about. I was glad to wake from my slumber, for I had had another nightmare, this time of us capturing and releasing Mohammed Habib.

"What's going on?" I demanded.

"Sir, Swampboy just killed a viper outside your door!"

"What?" I looked down to see grinning face of PFC Chris Blanc, aka. Swampboy. Swampboy was a Cajun from the bayous of southern Louisiana. He wore the grin of a Cheshire cat, and was stroking the body of a headless, yard-long, black snake. As he stroke it, the dead carcass twitched in an ethereal fashion.

"Yup, sir, got'im good an' dead. He wuz tryin' to get up in yo hooch, sir, but I saw'im and wacked hez ol' head clean off 'fo he knew wuz goin' on."

Leave it to a Cajun to kill a snake...and pet with the dead body afterwards.

"Well, uh, thanks, Private Blanc."

"Yez, sir, I learned long ago, you gotta get'im juz right in the neck - wack theya head clean off fast. You see, sir, I use'ta chase dem cotton-mouz down in da swamp. Dem cotton-mouz, you know, they bustle togetha in nests.

Even da younginz ah dangerous. So, you juz gotta go in and get the ho' mess of 'em. Wade into da swamp, go to daya nest, wack all daya heads off clean. If you don, they'll come back to bite you, sho'nuff. Maybe not today or tomorrow, but de next day for sho."

I stood blinking at the writhing body of the dead snake. Swampboy's touch stimulated the electrical nerve endings in its body, even though it lacked its head.

Swampboy continued. "Yes, sir, dem Afghan vipers are tricky creatures, too. Dey like to hide in places you don't expect'em."

A though was forming in my head. The young Cajun spoke the wisdom of a country-boy.

"Private Blanc, just what are you *doing* with that dead part of the snake and what do you *plan* to do with it?"

"Oh dis? Oh, shoot, sir, I'm just funnin' with its deadness. Soon as it calms down, I'm gonna throw it in the fire, head, body and all! That's the only thing dem vipers deserve!"

Cut off the head; burn body. Counterinsurgency 201.

Folding my arms, I replied, "I like your style, Blanc...I like your style."

I was tired of the Afghan vipers. Not just the snake variety, but the Taliban, ISIS, al Qadia, Laskar-e-Tayyiba, Harkat-ul-Jihad-al-Islami 313, We-worship-Allah-and-Want-to-kill-Everyone type of vipers that infested my world!

I got on the satellite phone and called my battalion TOC's S2 shop back at Camp Blessing. I told the S2, Captain Jason Bingo, I wanted him at my FOB the next day with all the current intelligence he had on the insurgent forces in the Watapur Valley. He resisted at first (I think he enjoyed the modest creatures comfort he had created within his own S2 shop), but when I sternly threatened to ring his neck and reminded him *he* was supposed to support *my* element (not the other way around), he relented. I planned to have an intel sync meeting - to synchronize what he was seeing, my meager intel men were seeing, and what MSG Jarrett's intel boys were seeing. I wanted a common operating picture (COP) of all the enemy's dispositions. Often the army tried to bury a COP in about 50+ PowerPoint slides. This didn't get the job done. It had to be done the old fashioned way - face to face, mano y mano.

The next day, I opened the meeting with a little spiel.

"Okay, I want all of you brainiac intel geeks to sit here together and get on the same sheet of music. I need everything you know about the enemy, down to the last detail. After we get a real common understanding of the enemy's dispositions, we're going to develop a plan to crush him."

MSG Cody Jarrett grinned and spat a long spittle of naswar tobacco into a cup. He knew I meant business.

CPT Bingo, however, blinked in disbelief. The S2 said, "We can't develop a battle plan that fast..."

"Oh, yes we can, and yes we will." I resolutely proclaimed.

It took several hours, but eventually we developed a pretty thorough understanding of what was going on in the Watapur and the enemy territory northwest of the Watapur, in the wilds of southern Nuristan Province. It's amazing how different things appeared when we shared information face to face. The main village in the western side of the Watapur was Chamadin, controlled mainly by Mohammed Habib. From there, going northwest one the border of Konar and Nuristan Provinces were Qaylaygal, controlled by Falaq Naz, Kwond Kalay, controlled by Nur Gul, and Mandesh Village, controlled by Abdul Haq. Nur Akbar floated between all these villages, but had less familial relations the further northwest he roamed, away from the heart of the Watapur. The location of the al Qaida Arab ghost, Barak al Qahtani, was anyone's guess. We did learn, however, that al Qahtani often was accompanied by another Arab trainer, an Egyptian named Muqtar al Masri.

Taking it all in, studying the map, I asked, "Whose the lowest hanging fruit? Who's daily habits and pattern of life do we know best?"

Muffled sidebar discussions ensued. Eventually a majority of the brain power came to the conclusion that Mohammed Habib was the best known.

Jarrett piped up, "Cap'n, I've got a man in Chamadin who can trigger an op against Mo Habib. We know where his compound is."

"Great! What else?"

Captain Bingo sheepishly raised his hand with a suggestion. "Well, I believe we can track Falaq Naz' movements."

"Really, how?"

"I can get Division HQ to chop us a Predator UAV to view his compound. He's one of the few men easily distinguishable from the air because he's only about 4'10" and has a 12 inch red beard. No mistaking him."

"Great! We'll use a Hellfire missile from a Predator ("Pred") MQ-1 UAV to knock out Falak Naz and M982 Excalibur rounds to take out Mohammed Habib."

Jarrett spat out his naswar and replaced it with a pinch of Copenhagen, stretching out his long legs in front of him. "Cap'n just one thing - have you thought about collateral damage? Are you willing to take the heat from the brass? It" be a helluv a lot cleaner for me to just pay one of my spies to assassinate them"

"Cody," I replied, "I don't give a *damn* about collateral damage!"

With a plan in place and (finally) some support from Division HQ (with the Pred), we waited for our moment. We wanted Jarrett's spook in Chamadin to give a signal via smoke (from burning donkey dung, no doubt) to tell us when Mohammed Habib was home. But, we wanted a simultaneous strike, so we had to keep multiple Pred's circling overhead waiting to see the "hobbit" Falak Naz in the open. This took patience; lots of tactical patience. I don't normally smoke, but I went through a pack of Marlborough Reds waiting to be Rikki-Tikki-Tavi - the mongoose who killed the vipers.

Finally, the day arrived and all the stars were aligned just right for us! Falak Naz was spotted walking away from his house with a group of his other armed minions just as it was confirmed that Mohammed Habib was at home. The artillery section located at FOB Wright to our south had their 155MM cannons loaded with M982 Excalibur rounds. The Excalibur rounds, though they cost about $60,000 a pop, were GPS guided precision artillery munitions. The coordinates for Mohammed Habib's compound were dialed in, and we had eight rounds ready to level his compound.

Receiving approval from higher, I gave the command to fire on Falak Naz and Mohammed Habib simultaneously. From the Predator feed, I saw the image of Falak Naz in the crosshairs. PFWOOM! A flash of light and dust exploded where the crosshairs had been, and one snake was dead. My FDC (Fire Direction Center) confirmed they fired all eight rounds at Mohammed Habib's compound. The Predator drone flew over the compound so we could witness the remains. The video feed from the Pred clearly showed the compound was obliterated!

"Now that's the way to do it men!" I shouted, proud that I did not have to risk a single American soldier to go and kill these evil men.

Days passed. We did not receive a single bit of sniper fire nor were any of our patrols or the ANA patrols fired upon. Even the stealthy IED's went silent. Then the intel reports came rolling in. They were not what anyone expected.

First I received a call from CPT Bingo. "Jack, its me, Jason."

"What's up?"

"Well, we are getting reports that Kwond Kalay is in an uproar over the death of Falak Naz. It sounds like his entire clan is taking up arms and planning some kind of retribution. In fact, they believe their village is infiltrated with spies and they are chopping off the heads of several innocent civilians."

"Why? We didn't have any collateral damage. And, from what you and the SF boys have indicated, we have no spies in that village."

"Hey, Jack, I'm just telling you what's going on - I can't explain the *why*."

My next call was from MSG Jarrett's chief 18F, his intelligence NCO, SFC "Madman" Marion Morrison.

"Hey, Captain Jack – it's me, Madman Morrison."

"What's up, sergeant?"

"Good news and bad news. My spy network and the Arbakai's are all a-buzz about the bombing of Mohammed Habib's compound. Sounds like we waxed the old bastard, but, we also killed his two wives and eight of his children."

I closed my eyes and prayed this wouldn't be leaked to the press. If it got out, my skull would be opened and the lawyers in Bagram would feast on my brains for dessert.

"Okay. What's the good news?" I asked.

"Well, sir, the supporters of the Taliban are being run out of Chamadin by the villagers. They believe the *presence* of the Islamic terrorists will end up getting more innocent lives killed. They're not mad at *us*, they're angry at their own Islamo-fascist leaders! As a consequence, they had a *jirga* and elected a new village elder and mullah for their village who has sworn his loyalty to the Afghan government and taken sides with the Arbakai!"

I hung up the phone and thought about these new reports. These were the *opposite* results of what I had been taught to expect about counter-insurgency at the Infantry Officers Career Course back at Benning or read in portions of FM 3-24. Here, I took pains to *limit* the collateral damage in Kwond Kalay, and the village turned *against* me instead of cheering for me. Likewise, I had given orders which resulted in the death of innocent lives in Chamadin and the villagers dumped their support for the enemy and were now supporting our side. Why? Stunned, I meandered outside to find a quiet place to have a smoke.

In a rather secluded corner of the FOB, I was enjoying a long drag of my Marlboro cigarette while sitting on a sandbag and facing the HESCO wall. Inhaling the sweet burning tobacco, I began to think very deeply about the bizarre nature of this war. Were all counter-insurgencies so illogical?

Suddenly, I could hear the crunch of the FOB's gravel-strewn ground crunch louder and louder with ever increasing intervals to my rear. "Ouch!" I exclaimed. A sharp needle-like pain jolted my left hand. I withdrew it quickly. An intense pain surged though my arm. I began to turn, my cigarette falling from my mouth, and saw a man lunging towards me with a hoe! I whipped around, stumbled, and fell off of the sandbag and onto my back. It was Swampboy!

WHACK!

The hoe had swung down towards my feet. Had the Cajun gone mad?!

Swamboy shouted, "I got 'em, sir, I got 'em! Cut that dad-gum viper's head clean off! I thought they'd be a nest of 'em back heeah in da corner o'da FOB. Damn Afghan vipers! The only way to get 'em sir, is to kill *all* of 'em! *Every last one of 'em*!

My head felt faint. I looked down and saw my hand swelling to a grotesquely large size. My breath became shallow and my heart skipped a beat. Woozy, I stammered,

"Private Blanc,Blanc...uh, call the medic quick! I fear you were too late. That damn viper got me first!" Then everything went pitch black.

Lessons Learned:

- *Leave PowerPoint for the generals. The only real way to plan is hack it out on a personal level.*
- *If you're not willing to fall on your sword for your decisions and your men, you shouldn't be allowed to ever raise a sword. Go find another profession.*
- *How much did that UAV cost for 3 days of overhead coverage? How about that M982 Excalibur round? Guess what - a single bullet from a paid assassin is a heck of a lot cheaper!*

- *"Collateral damage" can work in your favor if you properly target the snakes' lair; the enemy may be jealous of accuracy but respects and fears brute strength.*
- *In many third world countries, women are viewed as cattle and children, mere byproducts of intercourse. If the enemy cries about 'collateral damage' it's not because they really care, it's because they know* your *government,* your *media and* your *politicians will care.*

- *It's time for politicians to stop worrying about "collateral damage." If the enemy sees they don't care, they will fear you.*

- *The only good viper is a dead viper*

MAP 7 – EASTERN WAYGAL VALLEY, KONAR & NURISTAN PROVINCES, AFGHANISTAN
Waygal Village
PEAK 10,914 ft.
RIDGELINE
RIDGELINE
PEAK 9,129 ft.
OBJECTIVE SWORD
Mandesh
6,788 ft.
OBJECTIVE JUNO
Kwond Kalay
6,901 ft.
DIRECTION TO
WATAPUR VALLEY
OBJECTIVE GOLD
Qaylaygal
5,785 ft.
WAYGAL VALLEY
4,743 ft.
4712 ft
2011
Image © 2014 DigitalGlobe
© 2014 Cnes/Spot Image
Google earth
Earth Point
Imagery Date: 7/24/2010 35°04'48.63" N 70°56'58.55" E elev 8639 ft eye alt 23957 ft

CHAPTER 8 - Savagely Smart

I awoke thankful to be alive and with the epiphany that something had changed. Warfare had *evolved.* The expectations that had been instilled in me from time I was first commissioned were passé. Finally, after a more than a decade of fighting Islamic terrorists on battlefields, the generals and theorists had likewise concluded that the West was no longer pushing the evolution of warfare. The savages were. And, it became clear, the only way to kill the savages was to be more "savagely smart" - *more* cunning and *more* ruthless. There was an elegance to ruthlessness because ruthlessness worked. "Proportional response" to an enemy's action was idiotic. Overwhelming firepower ruled the day. Love and kindness did not win over your enemies; *fear* won over your enemies. And I had finally determined that I would strike fear into the hearts of my enemy and completely overwhelm them by any means necessary to obtain unconditional victory.

The turning of Chamadin village had been an unexpectedly positive tactical victory, giving us a tenuous dominance of the Watapur Valley. But, the killing of Falaq Naz in Kwond Kalay was only a pin-prick. I knew now from experience that the insurgents could send in reinforcements from the northwest if I didn't plunge into the snake-pits and kill all the vipers. Logically and logistically, the best solution was to destroy one nest at at time, starting in Qaylaygal and working north towards

Kwond Kalay and, finally, Mandesh. I wanted to brainstorm for ideas. So, again I turned to my Green Beret friend, MSG Cody Jarrett.

"Jarrett," I said, "We need to finish this. Let's get dirty and dive into each of these last three snake pits. Do you have any original ideas?"

"As a matter of fact, I do, Cap'n," he confidently replied. "I just received authorization to execute a plan we've had in the works for over a year."

Stunned at this news, I exclaimed, "Really? Well, what is it - do tell, please."

He looked at me askance. "Sir, this is really, really out of the box. I'm not sure I want to give you the details. It's a black op. Operation Gold."

Perplexed, I furrowed my brow. "So, you can't share with me any of the details because I'm not on a need-to-know basis, right?"

"Right. Just give me ten days, and I believe we can take Qaylaygal without firing a shot. If it doesn't work, then we'll go in with overwhelming force."

I pondered it for a second. I did not like being left in the dark, but I knew black ops were black ops for a reason.

"Jarrett, can't you even give me a hint?"

"Just remember Romeo and Juliet."

Now I was even more perplexed and intrigued. Although impatient for dramatic action, I nevertheless approved. "Okay, Jarrett, go ahead with your Shakespearean drama and let me know afterwards."

He gave me a maniacal grin and said, "Consider it done!"

Qaylaygal - Operation Gold

The silence was deafening. Eight days had passed and there was almost no activity on the battlefield. I have no idea how many dusty, dog-eared *Maxim* and *Hot Rod* magazines I flipped through waiting to learn the results of Jarrett's plan. It was driving me crazy, but I was not entirely idle. I devised plans for a large assault into Qaylaygal Village . There were reports that, ever since our night-time assault against Gambir village, the enemy had been posting sentries at night, waiting for us to attack in the dark. They had become so accustomed to our night raids that intelligence indicated they were now *more* prepared at night than during the day. Thus, my plan was to conduct a bold daytime raid. But, this assault would be anything but normal. I procured 50 *sharwal kameez* "man-jams," pakol hats, and vests for my men from Jalalabad. My plan was to dress up part of my company in local Islamic garb as a ruse, and use another part of my company, in normal US uniforms, as bait. There were multiple, solid intel reports that enemy reinforcements were headed south into the Waygal District from the northern Nuristan district of

Kamdesh. Thus, I figured we could gain an element of surprise if the insurgents thought, even for a moment, that my men were *their* reinforcements.

The silence was suddenly shattered by staticky chatter from the radio.

"Sir, this is the main gate. You're not going to believe this, but we just captured one of the main insurgents right outside our gate!"

"What?!" I exclaimed. "Wait right there; I'm on my way."

When I arrived at the gate, I was surprised to recognize a face I had seen in old Taliban propaganda videos. The man had been zip-tied with flex-cuffs behind his back. It was none other than Mullah Hakim Gul, the recidivist Islamic cult leader of Qaylaygal!

"Sir, this is Mullah Hakim Gul!" the gate guard proclaimed.

"I know, I know. What happened."

"Sir, he just walked up to the gate unarmed and said he was here to make peace. He said the villagers and clans of Qaylagal wanted to be integrated into the Afghan government and become part of the Arbakai!"

I strongly suspected a ruse. Mullah Hakim Gul as an infamous, radical Salafist cleric. He had been captured years earlier and imprisoned in Bagram for over a year, before tragically escaping the prison and disappearing into the wilds of southern Nuristan. We knew he was the main cleric in Qaylaygal, but little else.

A felt a strong hand gripped my shoulder from behind. I turned to see MSG Jarrett.

"It's okay, Cap'n. I know you think this is probably a ruse, but come with me and I'll explain."

I gave orders to un-cuff the mullah and have him detained in our solitary holding cell. I followed Jarrett back to his TOC.

"What's going on, Jarrett?!" I exclaimed.

"Qaylaygal is ours."

"What?! How?!"

"They're all dead."

"Who's all dead?"

"All the insurgent leaders in the village."

"What?! How?! I haven't heard a shot fired."

"Exactly. Operation Gold worked perfectly."

Still shocked at this, I sat down and "borrowed" a pinch of Jarrett's fresh naswar. "Explain."

"Cap'n, over two years ago, Mullah Gul was captured in Asadabad. It was a lucky capture, but a positive one nonetheless. Knowing he was a virulent Salafist, my command decided to try something new. Re-education. Re-programming. Propaganda at its finest.

"Contemporary Islam believes there are five pillars of their faith. Some sects add a sixth - jihad. Jihad - which means "struggle" - can be interpreted two ways: as a violent, physical fight against anyone who is not a Muslim or as a personal, internal struggle to be free from sin.

"When Mullah Gul was in prison, we planted his cell with clerics who believed in the non-violent form of jihad. These planted clerics studied the Koran with the mullah every day for hours upon hours. Eventually, they convinced Mullah Gul to renounce his evil version of jihad - but to keep his renouncement a secret. He never knew it, but these planted "prisoners" were actually American-Afghan clerics from California! After we brain-washed him, we recruited him as one of our spies."

My eyes widened at the genius of this ruse. We used their own religion against them to re-brainwash them! Amazing.

"So, he didn't really escape from prison, did he?" I inquired.

"No. We only made it seem as if he escaped. Another ruse."

This didn't complete the story for me, however. I needed to know more. "So, great. But, you said 'they're all dead.' What did you mean by that."

Jarrett continued to explain Operation Gold to me. "Since Hakim Gul is a mullah, all the insurgents look to him for divine guidance. He called a *jirga* meeting in the village at their mosque for all the clan insurgent leaders. Then he poisoned them."

My eyes shot wide. "He *what*?!"

"Mullah Gul spiked their tea with a poison we concocted that would kill them off in their sleep within 24-48 hours. When they all died in their sleep, Mullah Gul called another *jirga* meeting at the mosque and declared Allah told him in a vision the insurgent commanders died because they lived a life counter to the will of Allah, and that Allah willed for the people of Qaylaygal to reconcile with and support the Afghan government and American 'people of the Book.'"

"And the villagers bought it?"

"You might say they 'drank the Kool-Aid.' Some of these primitive cultures strongly believe in the supernatural. Why not use that against them?" Jarrett explained.

"So now what? What's next?"

"Cap'n, I need you to get those Preds from Division in action ASAP."

"Why?"

"Because, Cap'n, the bodies of the bad guys are on their way to Kwond Kalay by donkey for a mass funeral tomorrow."

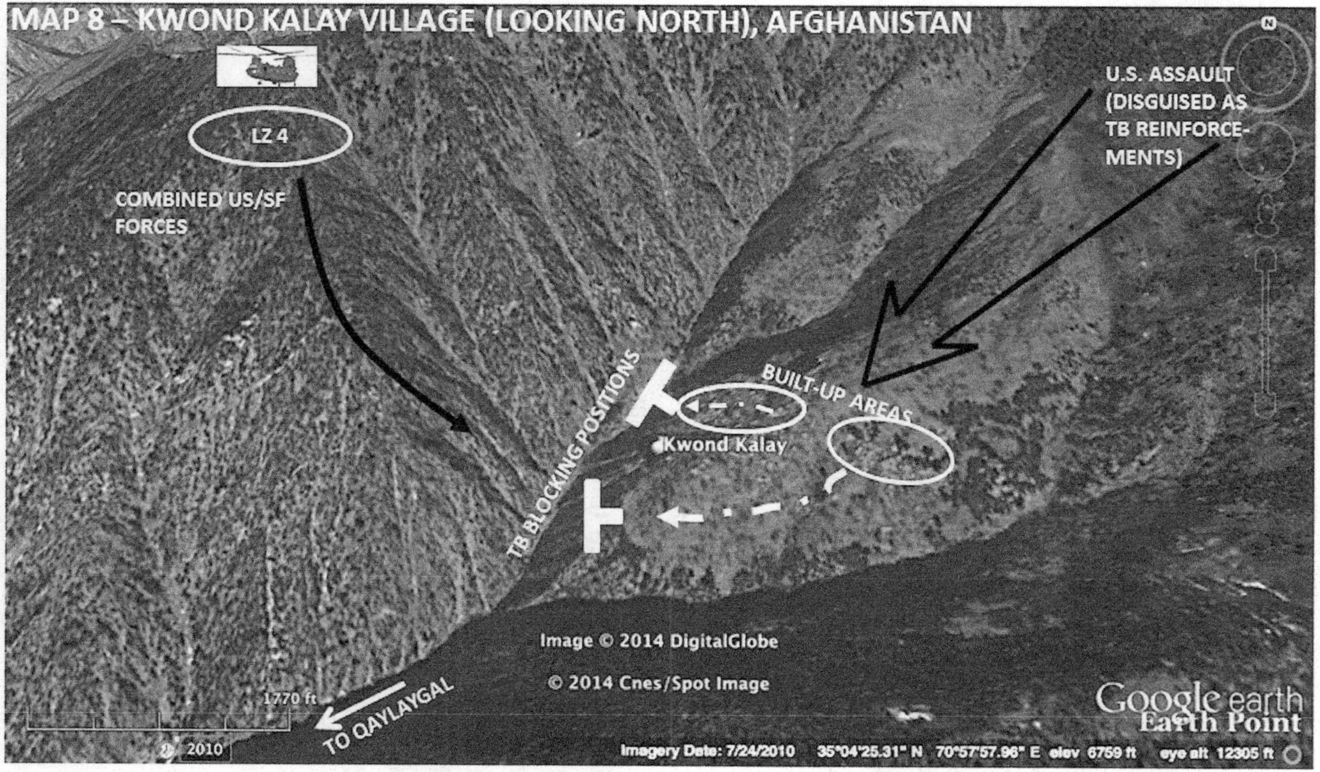
MAP 8 – KWOND KALAY VILLAGE (LOOKING NORTH), AFGHANISTAN
LZ 4
COMBINED US/SF FORCES
U.S. ASSAULT (DISGUISED AS TB REINFORCE-MENTS)
BUILT-UP AREAS
Kwond Kalay
TB BLOCKING POSITIONS
TO QAYLAYGAL
1770 ft
2010
Image © 2014 DigitalGlobe
© 2014 Cnes/Spot Image
Google earth
Earth Point
Imagery Date: 7/24/2010 35°04'25.31" N 70°57'57.96" E elev 6759 ft eye alt 12305 ft

Kwond Kalay - Operation Juno

Thankfully, Division HQ still had multiple Pred's dedicated to our area of operations following the successful killing of Falaq Naz. Jarrett and I conceived and planned Operation Juno in about two hours. We had to act fast, lest some bureaucrat decided to put the kit gloves back on. Since we had been released of most of our self-generated ROEs, actions similar to what we had done were being executed all over the theater. (Perhaps the ISAF commander had sent the lawyers on a paid RnR vacation to Hawaii.)

It always amazed me that there seemed to be a shortage of helicopter lift assets in Afghanistan. I never quite understood this. It was as if the brass didn't want to be fully committed to the war effort. But, as I mentioned, things were changing. The "geniuses" at the Pentagon had finally come to the realization that we needed more helicopters in theater. So, they quickly mobilized every National Guard aviation unit to join us in Afghanistan and flood the third dimension of the battlespace - air - the one place the enemy had zero presence in. (Gee, it only took them 12 years to figure that out!)

I was seated comfortably in my new Afghan man-jams in my TOC gripping my borrowed AK-47. I was proud the half of my company that wore the local attire had also learned to effectively use the AK quickly. I had been watching the MQ-1's video feed, following about eight donkeys as they marched up-hill from Qaylaygal to Kwond Kalay. The struggling creatures appeared to have bundled, blanket clad dead bodies roped to their back. The video was grainy, but the bundles were definitely not food-supplies or sticks.

Finally, the bodies were unloaded on the outskirts of Kwond Kalay. A large group of men gathered around the dead bodies, their lifeless faces exposed and heads circled with flowers. Funerals were often used in propaganda videos to extol the virtues of martyrs. And anyone who attended a funeral of a scumbag was a scumbag.

I was on an open line to division who was talking to a Pred controller who was based far, far away, outside of Las Vegas, Nevada.

"Targets identified," the VOIP line crackled.

"FIRE." I commanded with resolve.

Moments later, the screen flashed as multiple missiles slammed into the group of funeral mourners. Bodies flew in the air like rag dolls. Some of the wounded dragged their legless bodies across the rocks. More missiles slammed home, mercifully killing the wounded. The Pred camera scanned furiously for stragglers. None were left.

"Okay, boys, that's Phase I of Juno. In 24 hours we fly!"

We knew the killing of insurgents at a funeral would flare up a hornet's nest and draw in enemy reinforcements. But we would get there before them. We had helicopters and would appear during the daylight hours, when they least expected it.

I wanted my men totally pumped up for this operation. We cranked up the loudspeakers and filled the night air with heavy metal music. *Metallica, Slayer, Megadeath, Ramstein, Slipknot* - we played them all. But, it was not only my men that I wanted to get fired up. I wanted the locals to know something was about to happen.

I wanted them to hear the madness of our music and *fear* what was going to happen. I wanted them to fear the unknown; to quake at the threat of our wrath; to shiver at the hunter who stalks the night. But, this time, our crusade would come in the daylight!

The man-jam clad portion of my company actually flew out that night. After several false insertions, we dropped off in a very remote location north of Kwond Kalay. This was the direction from which the enemy would expect the reinforcements to come. We carefully chose local and loyal Kamdeshi Arbakai's as our guides. They would explain we were Islamic reinforcements to any innocent shepherds who we might encounter. Each soldier carried two pakols. One pakol was either tan, white, or red in color and appeared to be of Kamdeshi origin. The second was all black. We would all don the black pakol ar prior to our assault and help us identify friend from foe. After marching all night, we were 1KM outside of Kwond Kalay at sun-up. This would normally be a very difficult task, but we had the advantage of moonlight, and were not encumbered by bulky body armor. (Now I understood how quickly the enemy could move.) I had the guides lead our men in the prostrating motion of morning Islamic prayers to make the ruse appear legitimate. By my side was I had three men: PVT "Snakeboy" Blanc, Green Beret SFC "Madman" Morrison, and a plucky US Air Force JTAC (Joint Terminal Air Controller) named Technical Sergeant Charles LeMay IV. Madman would help communicate with MSG Jarrett's element who, acting as bait, was going to attack from the south with his Afghan Commandos and the remainder of my company. TSgt LeMay's duty was to call in Tactical Air Support if things went sideways.

Our Arbakai guides were warmly embraced by the tired insurgent sentries on the outskirts of the village. The insurgents recognized the guides' Kamdeshi dialects and were convinced we were the reinforcements they called following the funeral massacre.

"Praise Allah, our Brothers. We have been awake all night expecting an American attack after the disaster at the funeral yesterday. Please join us for some morning tea."

"Yes, thank you my Brothers. We have heard of your struggles recently over the ICOM radio and marched all night to fight by your side." the guide replied. "Make for us some more tea and I will retrieve my commander and the rest of our men."

Our Arbakai guides began marching towards us with the sentries at their side. Like clockwork, the distant sound of Chinooks whipping the morning air echoed in the valley to the south.

Alarms sounded in the village. Shouts and calls for men to arms. Insurgents spilled from their mountain hovels like ants erupting from an overturned ant hill. I had no idea there were so many armed men in Kwond Kalay! Intelligence reports had greatly underestimated their number. Other reinforcements must have arrived *prior* to our strike on the funeral.

"Brothers, brothers, come quick! You are just in time! The infidels are arriving!" shouted the sentry to one of my guides. They ran south to take up positions to intercept the direction our conventional element would approach.

My men donned their black pakols and released the hounds of Hell!

I guesstimate that we took out about 50% of the insurgents by shooting them in the back. That was the easy part. Sweeping through the village, bounding uphill and downhill, fighting from room to room, was exhausting. We were careful not to kill any women or children, except for those who took up arms against us. (Several women and boys charged with grenades.) Since we were carrying a light load of AK-47 ammunition, some of my men ran dangerously low. Some, like Snakeboy, squandered his 7.62 x 35mm ammo by firing on full-auto, and had to resort to other means. I saw him take a strand of Islamic prayer beads and strangle one of the desperate Taliban fighters to death. I filed that scene away in my head, and decided to recommend him for a medal for ingenuity.

The Battle for Kwond Kalay was over in about an hour. We took only three prisoners. Two were local Afghans Taliban fighters, but the third was particularly important. It was Muqtar al Masri, the Egyptian right-hand man of the evil Saudi, Barak al Qahtani!

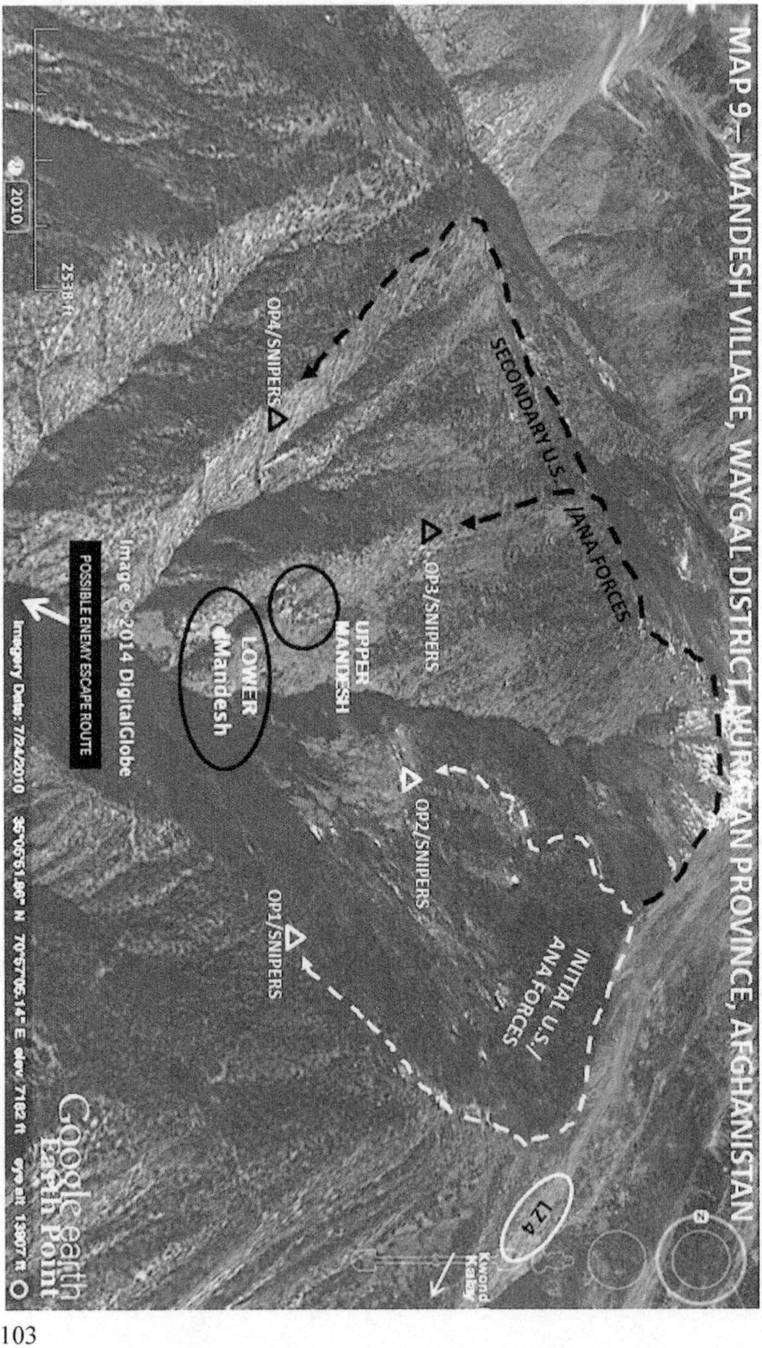
MAP 9 – MANDESH VILLAGE, WAYGAL DISTRICT, NURISTAN PROVINCE, AFGHANISTAN
SECONDARY U.S./ANA FORCES
OP4/SNIPERS
OP3/SNIPERS
UPPER MANDESH
LOWER Mandesh
OP2/SNIPERS
OP1/SNIPERS
INITIAL U.S./ANA FORCES
LZ4
Kwond Kalay
POSSIBLE ENEMY ESCAPE ROUTE
Image © 2014 DigitalGlobe
Google earth
2538 ft
2010

Mandesh - Operation Sword

"Where the hell is Barak al Qahtani!" MSG Jarrett shouted in Arabic to Muqtar al Masri. Al Masri had been captured while attempting to flee the battle in a woman's burqa and blue dress. In the mountains for southern Nuristan, however, women did not wear burqas because they too cumbersome for work in the mountains. That was first our tip-off. The second were his size-12 feet.

Al Masri had been taken outside the village to a secluded area strewn with large boulders and free from prying eyes.The only men present were Jarrett, one of the

Commando interpreters, and the Afghan Commando commander, Captain Zubair, and me.

"I know you understand me, son-of-a-pig!" shouted Jarrett in Arabic again. "Tell me where the hell Barak is."

Al Masri turned his head and refused to look Jarrett in the eyes.

Jarrett quickly withdrew a K-bar knife and held it under Al Masri's groin. "Tell me where Barak is or I'll cut your balls off!!" Jarrett barked in Arabic. That got his attention. Suddenly, the seasoned al Qaida operative looked Jarrett in the eyes and showed a glimmer of fear. He was trained to resist interrogation and knew the US Interrogation Manual by heart. He knew the rules by heart what we could legally do to him. Yet, we had changed the *pattern* of how we fought by assaulting in daylight wearing local attire. The Egyptian Islamist feared the rules had changed.

"Pin him to the ground. Hold his legs. If he doesn't have the balls to tell me where Barak is, he doesn't deserve to have any balls," shouted Jarrett. Jarrett cut a hole in al Masri's man-jams exposing his testicles.

Zubair gently grabbed Jarrett by the arm. "Hold it, hold it, my brother Jarrett, let us handle this. We'll get the information out of him...the Afghan way."

I won't tip any TTPs[8] by explaining the painful methods Zubair used to extract information from al Masri. But, it was clear Zubair had trained and practiced the art of

---

[8] Tactics, Techniques, and Procedures; in other words, how things are done.

extracting information through unconventional methods. Within an hour, not only did al Masri talk, but he pointed out for us, on photographic imagery of Mandesh village, the precise location of al Qahtani. He also revealed that Mandesh had been flooded with insurgents from all over the region. There were 32 buildings in Mandesh village. Four of them were used as centers for al Qaida, HUJI 313, and other operatives to train the Taliban in *international* terrorism. Al Masri even confessed a new ISIS cell had recently formed there. Mandesh was a ticking time bomb that we wouldn't have discovered without these enhanced interrogation tactics.

"Okay, Cap'n. Better get back to your company."

"On to Mandesh?" I asked.

"On to Mandesh."

I walked away, my brain filled with the vital and actionable intel al Masri had given us.

POP!

I heard a 9mm pistol fire from where al Masri was being held. Alarmed, I ran back to the location. I saw al Masri's body in a pool of his own blood and excrement, a small red hole between his eyes. Jarrett held a smoking pistol.

"What happened?!" I exclaimed.

Jarrett had a wide, mischievous grin on his face. He declared, "Scumbag tried to escape."

We knew from listening to the enemy ICOM radio chatter that an alert had been broadcast to commanders in or near Mandesh that Americans dressed as local civilians had attacked Kwond Kalay. From the radio intercepts, it was clear that enemy paranoia was erupting to a fever pitch. Now, the terrorist center at Mandesh had no idea what to expect. Would we attack in local garb or American; during the night or during the day? They were scrambling to devise an effective means to fight. Some suggested that they flee further north, deeper into Nuristan. Others demanded that everyone stand their ground and fight. We had moved so quickly and changed our tactics so rapidly we were now truly "inside the enemy's decision cycle."

Everywhere things appeared bleak for the enemy. They requested reinforcements from the south, across the Pech river in the Korengal Valley, but the radio intercepts indicated *my* higher command had ordered two battalions of US Marines to sweep through that dreaded valley and decimate the enemy. (It was odd that I found this out from the enemy *before* my own higher chain of command told me, but that's not unusual in the fog of war.) There were also cryptic ICOM radio chatter about "death phones" - cell phones that blew up when insurgent commanders used them. This story was retold by one of our Afghan captives.

(It was only after I returned home from the war that I learned the CIA and NSA had collaborated with the Israelis on a computer virus that corrupted specific cell phones, positively identified insurgent commanders from voice recognition software, and then triggered tiny explosives built into the cell phones, putting a nice hole in their head.[9])

Again, I changed my tactics. This time, I decided to slowly squeeze the enemy by surrounding Mandesh Village with sniper fire. First, we would find and eliminate their sentries and block their likely exfiltration routes. Then, we would tighten the noose, until Mandesh could become a shooting gallery. Now, my scout snipers had not had as much practice in combat as those who fought in the urban terrain of Iraq, but they could be just as effectively lethal in this situation. But, we had to move quickly and keep the enemy off balance, since they were still uncertain what steps to take. We also knew Barak al Qahtani and his Arab cells would be the first to flee. We were in a race against al Qahtani as much as anything else.

Because my man-jam clad element was traveling so light, they needed an emergency resupply of ammo before our hump to Mandesh. But, the Commandos, the SF team, and my "regular" troops were still fresh and fully supplied.

---

9 Note: I totally made this up, but wouldn't that be a damn sexy program! I admit I was inspired by the Israeli killing of Hamas bomb-making expert Yahya Ayyash in 1996. We could learn a lot from the Israelis.

Giving them the intel data dump we gained from al Masri, I immediately sent them forward by foot. To guide them, I gave them the local, Watapur-based Arbakai guides. I also gave them control of our two Kwond Kalay Afghan captives as guides. (They become strangely motivated to show us the enemy routes *after* we showed them the oozing body of Muqtar al Masri, now covered with flies.) Wanting to lead from the front, I joined my fresh, regular troops, along with the Air Force JTAC.

Though Mandesh was only two kilometers from Kwond Kalay as the crow flies, our route to high-ground around the village was indirect, and the going was slower than I liked. So, I ordered my scout snipers and my most physically fit platoon to take the armored plates out of their carrier vests and march like the wind. Without the plates, and not wanting to be bested by local guides wearing mere sandals, they turned into a version of Stonewall Jackson's modern "foot cavalry." It was not long before they gained contact with the enemy. The insurgents had already sent out patrols to feel out our disposition. The snipers made short work of them, but we weren't certain if they had raised an alarm before going to meet their 72 virgins in Paradise. Gaining a toehold on the highest peak we could find, I radioed back our location and told the rest of my company to move ASAP, even if it meant traversing the mountains in the dark.

At first light, we had secured three of the four hills around Mandesh. This was, in large part, due to the Afghan

Commandos who slithered up behind the hill top Taliban sentries and quietly slit their throats with their Khyber knives.

Because of the geography, our snipers could not completely surrounded Mandesh. However, perched on high-ground and using crosstalk over a dedicated radio frequency, the sniper teams described in detail their coverage and dead-spaces. They believed they could cover 50% of the village. After sunrise the next day, the village appeared eerily deserted. But, we knew it was not. Tell-tale signs of humanity were everywhere, from the animals, to the smoke from the stovepipes, to the smell of fresh baked flatbread.

At 0900, we had our first contact in Mandesh. A curious military aged male poked his head outside a window to see what was going on. He took a .308 bullet from a Remington 700 between the eyes. At 1015, this repeated itself.

"Sir, dis is gonna take forevah," confided PVT Blanc by my side.

"Okay, Snakeboy. If this was a den of snakes, what would you do?"

"Well, cotton-mouz, dey have de-ah nests in da water, so you can't smoke 'em out. But rattlers, well, by-gum, if don't mind starting a small fire, you can smoke them SOB's out by lighting a fire at one end of their hole."

"Okay, Snakeboy, good idea. We'll smoke the bastards out!"

I ordered my mortars to put a delayed fuze on their Willey-Pete (White Phosphorus) rounds and aim them at the buildings we believed were the al Qaida training compounds. With the beautiful sound of "whump...whump...whump" the rounds arced high in the air and landed amongst the buildings. Two structures caught fire. Ten minutes later, scores of Kalashnikov armed men came pouring out like ants on fire.

Our snipers went to work. They did the best they could given the dead space and moving targets.

One of our imbedded Afghan troops, monitoring the enemy radio traffic, came up to me with Madman Morrison.

"Hey, sir, I think we've got good news and bad news."

"Okay, Madman, let's have it."

"Our commandos picked up a bundle of radio traffic. They kept saying things like, 'We've got to protect the sheikh; we've got to get the sheikh and his men out of here.' The 'sheikh' has to be a reference to Barak al Qahtani. He's still in the village, I'll stake my reputation and trip to Leavenworth on it!"

"Okay, what's the bad-news."

"Sir, we tried to skirt the west side of the village. I don't think we can completely surround the village."

This was indeed bad. Maps don't tell you enough about the terrain. You never know the full lay of the land until you get there. We only had two days of supplies. Safely flying in additional supplies would take precious time - time enough for the enemy to slip beyond our grasp. I released a sign of remorse. I was stumped. The men around me saw it in my face. They knew the facts as well as I.

"Hey sir, I got an idea," chimed in TSgt LeMay.

"I'm open to ideas, LeMay...let's have it."

"Why don't we just bomb these savages back to the stone-age? We know what's in that village. These are the minions who are planning *international* terrorist ops..."

"...which, as a threat to the Homeland, makes them fair game!" I finished LeMay's sentence.

"Exactly. Heck sir, daylight bombing worked in WWII and the Persian Gulf War in '91 and it'll work just as good today. In fact, there's a squadron of B-52's in the air right now slaughtering bastards down in the Korengal right now!"

"Can you get them diverted here?" I inquired.

"Absolutely. Stand back and watch the awesome power of the United States Air Force!"

Using the latest Ground Laser Target Designator, the Northrop-Grumman, the GLTD III, TSgt LeMay aimed at the roof of the building that we believed was "Sheikh" Barak's compound. Then he called in the B-52's.

I didn't see the huge bombers, but I did see their contrails etching white lines high in the sky under the noon-day sun.

"...Roger, fire for effect, grid 42SXD7701385689. Bombs away! Impact in 30 seconds!" shouted LeMay.

“Better plug your ears, sir; we’re about to ‘go-Guernica’ on these savages!”

What I saw next will be freshly etched in my mind until the day I die. First, the main building in the center of Mandesh flashed orange and gray. Two second later, 30 laser guided 500lb bombs exploded lifting what was once Mandesh village 100ft in the air. The ground shook with such force, I was knocked on my back like an earthquake that registered 10 on the Richter scale. Billows of orange, yellow, and gray mushroom clouds filled the sky. It was truly awesome. It took sixty seconds before the dust began to clear. LeMay aimed his GLTD at the few remaining structures, called in for more bombs, and it happened all over again. This was war at its finest!

In less than five minutes, Mandesh village and more than 100 Afghan and Arab terrorists and terrorist families were all dead. And our war was over.

At long last, there was peace in my valley. In fact, peace sprang up throughout the land like beautiful flowers, flowers fertilized by brutal combat and savage tactics rather than limited warfare, half measures, and feel-good political correctness. Victory, real victory, was finally and definitively ours. Identical tactics were not used in every corner of the theater. There were different solutions based on different original ideas, different tribal dynamics, different Islamic insurgent groups, varied geography, and individual American ingenuity. However, we all did share one thing in common in order to defeat radical Islamic terrorism. We had all learned how to be *savagely smart*. We all learned the *elegance of ruthlessness*!

VICTORY!

Lessons Learned:

- *The first rule of warfare is there are no rules.*
- *The goal of warfare is to win. Do* whatever *it takes to win.*
- *You can't see inside the hearts and minds of the enemy or the populace. But, you CAN make them FEAR you.*

- *What was that that Patton & Napoleon said? "Audacity, Audacity, always Audacity!" (No wonder the Germans feared them.)*
- *Changing the rules (not just your tactics) keeps the enemy off balance.*

- *What's the difference between targeted killing (aka. assassination) by a Hellfire fired by a UAV at 20,000 ft. and targeted killing by a cup of poison? A cup of poison is more precise and costs much, much, much less.*
- *You know that "Box" they put your brain in when you were at FT Benning, Quantico, Leavenworth or Carlisle Barracks? Well, you better start to think outside of that box and put your thoughts into action because I guarantee your enemy is.*

- *Interrogation is an art, not a science. So is torture. Torture can work if employed by an artist and not a goon.*
- *As I mentioned in Ch. 6, Captured terrorists should be killed after all information is squeezed out of them.*

- *B-52's equal Beauty. Saturation bombing worked for grandpa in WWII and it'll still work for you!*
- *Neither whack-a-mole nor inkblots is enough. Overwhelming and annihilating the enemy produces peace and prevents the "crazies" from killing the innocents.*

PART II

Savagely Smart Theory

"You may not be interested in war, but war is very interested in you."

- Leon Trotsky

"These people do not want to negotiate...not at all. They want the universal caliphate established across the face of the earth and they want every infidel converted...or dead. If we take our foot off of the throat of this enemy for one minute...our world changes completely."

- Ed Hoffman, Body of Lies

The Theory of Relativity & Savagely Smart in Military Science

Relativity in Warfare

Warfare has been described as both an art and a science. It is an art inasmuch as the *creativity* that goes into designing a unique strategy, campaign, or tactics necessary to accomplish the goal: to obtain victory. It is a science in that there are certain principles and theories that stand the test of time. Like hard sciences, the theories that make up the body of knowledge behind its practice rest on theories which must (and should) be tested and retested; ideas which must be validated and revalidated. But as a science, the conventional wisdom (of a certain *era*) needs to be constantly challenged. How did Galileo, Newton, and Einstein advanced astronomy and physics? Only through rigorously *challenging* the conventional wisdom of *their* day and time. Therefore, I likewise encourage you to challenge the conventional wisdom of military science as it stands today.

When I was a young officer, we would always discuss "battlefield calculus." My first brigade commander was Colonel David D. McKiernan. (He went on to become *General* David D. McKiernan.) He was brilliant in "battlefield calculus" and drilled us on it constantly. In the field, he would give us pop quizzes, like "what's the standard time necessary to breach a complex obstacle given X number of plow tanks, Y number of engineer assets, and

Z minutes of smoke?" It was such in-depth calculus that let him to be one of the principles architects of the incredibly successful *Operation Cobra II*, which decimated Saddam Hussein's regular forces and led to the fall of Baghdad weeks after *Operation Iraqi Freedom* commenced. Battlefield calculus is indeed a very useful and *necessary* piece of military science. But, like Newtonian physics, it does not explain the whole science of warfare.

In Admiral William H. McRaven's brilliant (and under-read) work, "Spec Ops: Case Studies in Special Operations Warfare: Theory and Practice," we see the mind of the man who planned *Operation Neptune Spear*, the 2011 raid on Abbottabad, Pakistan that killed Osama bin Laden in 2011. In this work, McRaven postulated the theory of "Relative Superiority." In short, he postulated that a force of inferior size could defeat a much larger force if executed against the enemy's proper "point of vulnerability." *Relative* superiority relies largely on the principles of speed and surprise. This is not to say that McRaven discards the older principles of warfare: MOM-USE-SOS (mass, offensive, maneuver, unity of command, simplicity, economy of force, security, objective, surprise.) On the contrary; he built on these principles, refined them, and thus advanced the science behind the evolution of warfare.

Nonetheless, neither McKiernan's battlefield calculus nor McRaven's relative superiority explain the *relative* nature of human *reactions*. The human element is relative to both cultural norms in a given location and the erratic general nature of human beings. Therefore, there is never one right answer for how to conduct a military campaign. 1+1 *may not* = 2 against Iraqi *Fedayeen;* 1+1 *may* = 2.459 in Iraq; -0.46 in Afghanistan; or π (3.14159) in China. The best answer needs to be defined by what you want to accomplish. It should not be defined by political correctness or careerism or what makes you "feel good." Ultimately, the final outcome that you should *always* want to achieve its complete victory. Therefore it is from the perspective of total victory that you should construct and postulate your hypothesis on how to go about being successful at that aim.

Where we Westerners consistently miscalculate is embedded in our theory that human nature and human norms are constant throughout the planet. But they are not! There is *no* such thing as universal civil rights because all societies in all cultures are different! In physics, Einstein postulated the Theory of Relativity; time and space were relative depending on the position and acceleration of the observer. So it is thus with human reactions to a given force. Therefore attempting to construct a battle theory based on your perceived cultural norms is bound to fail because what you are essentially doing is attempting to put a square peg into a round hole. The best solution is not to force square peg into a round hole but to be amorphous and flexible enough to *rapidly* alter your battlefield calculus according to the environment in which you are waging war. For Einstein, in physics, the speed of light, *C*, is the one constant. In warfare, the one constant part of the equation is that *V* (victory) must be the the end result of *whatever* you *do* to achieve it. How savage that equation is to achieve *V* does not matter.

Savagely Smart

I recognize that many people will ask me for a good definition of the term "savagely smart?" Does this term imply that we should lock captured ISIS combatants in a cage and set them on fire as they do; should we rape women, pillage villages, and crucify Islamic soldiers as they do? Absolutely not! What I am trying to imply is there is a means to be *Savage* and to defeat savages without necessarily employing identical grotesque tactics they use. For example, Comanche and Sioux Indians were fabulous savage warriors, but they also horrendously tortured Western pioneers and soldiers, butchered their bodies, and frequently gang raped women. Were Quanah Parker and Sitting Bull successful? Absolutely; US troops struggled to combat their unconventional and savage tactics. In the case of the Comanches, how did Colonel Randall Slidel MacKenzie defeat Quanah Parker and the Comanches - an enemy that was nearly impossible to find and had superior horsemanship skills? MacKenzie's American soldiers *innovated*. After fruitlessly chasing the Comanches across west Texas, MacKenzie eventually found and butchered the Comanche's horse heard, thus cutting off their mode of transportation. For Sitting Bull and the Sioux, Americans *innovated*, killing the buffalo and thereby destroying their main source of sustenance. Were these tactics "savage?" Absolutely! But, I would define these tactics as *savagely smart*. They produced the desired results. These tactics were the relative calculus needed to achieve victory. A more contemporary example is the American experience in World War II. The Japanese fought with severe brutality. Should the American military have used suicidal *kamikaze* tactics in order to match that savage tactic employed by the Imperial Japanese? No. We created a different savagely smart method: the atomic bomb. In all the examples above, I would argue such savagely smart tactics employed by American forces ultimately saved the lives of more enemy combatants and noncombatants than employing tactics mirroring those the enemy used against Western forces. These savagely smart tactics also greatly shortened these wars and ultimately saved American lives, too!

Because human rights are relative rather than universal, and because there is no one common enemy, I propose that you innovate your own unique savagely smart tactics in order to achieve the ultimate desired results: complete Victory! America has faced many challenges and many different kinds of enemies. By employing savagely smart methods, America can defeat the threats posted by the Islamic State (ISIS), al Qaida, neo-Communists, narco-drug cartels, Russians, Chinese, and other current and future cultural, military, and economic enemies of the United States. America can and should once again *lead* the evolution of warfare.

Author's Note

I graduated from the United States Military Academy in 1993. During the last 21 years, I have served in a variety of unique capacities and in a variety of unique wars. I have served in three Balkan conflicts (Croatia/Bosnia 1996-1997 and Kosovo in 1999 and 2001, and Macedonia in 2001). I served as an advisor in Kuwait prior to Operation Iraqi Freedom (2002-2003) training young captains in tactics. Later I served extensively in eastern Afghanistan, doing short tours in 2007, 2009, 2010, 2011, 2012, not to mention two great adventures in west Africa (with ODA 014 1/10 SFG in Niger in 2007 and ODA 0236 2/10 SFG in Mauritania in 2008). In other words, I've been around the block a bit. Through it all, I've worked under some of the best commanders in the military (and probably some of the worst.)

I remember distinctly my first force-on-force exercise as a young butter-bar. It was my platoon against two of my peers. During the first mission, I gave orders for one of my radios to be put on the opposing platoon's radio frequency. Just before we attacked, I ordered a "jump plug" to be put in the radio that was set on the "enemy's" frequency, effectively jamming their platoon's radios at a critical time. For another platoon, I had some of my men dismount, infiltrate the opposing platoons assemble area. With chalk, we wrote crude comments on the sleeping "enemy's" tanks then woke them up by detonating Hoffman simulation rounds all around them. I pushed my men hard and our employed our machines in unorthodox places where they weren't designed to operate. I forced my men to train and maneuver at night. I broke the rules. And, I always won.

The other LT's thought I was a bit of a snot. "You're not fighting by the rules," they decried. But, I always believed that the first rule of warfare was there were no rules. The point was to win; a pretty simple premise.

In the early part of the 21st Century, America has fought a very formidable set of foes. They are formidable *not* because they are sophisticated, but because they don't fight by "the rules." There is nothing special about them. They are not supermen. But, we have failed to learn from our mistakes, learn from our enemy, and learn that warfare continues to evolve. Advanced technology does not evolve warfare; how you employ yourself and your forces evolve art warfare.

Plato wrote thousands of years ago, "Only the dead have seen the end of war." He was correct because, as Thucydides promulgated during the Peloponnesian War, wars arise out of fear, honor, and self-interest - key components of the nature of Fallen Man. What century you are in doesn't matter. What matters is learning with an open mind...and winning! I hope that you enjoyed this little book and that you read it (and re-read it) with an open mind. It might just save your life someday.

TSJ - April 2014

Thomas St. James' Recommended Reading List

The books below contain elemental threads of the theories displayed in *The Elegance of Ruthlessness.* I have listed these books alphabetically, for I do not want to give weight to one book more than the other.

American Sniper, by Chris Kyle. This is more than just a book about the life of America's best sniper ever and one of Texas' modern heroes, but it also contains critical elements about how we should fight on the battlefield to be victorious.

The Art of War, by Sun Tzu. This Chinese classic, written approximately 500 B.C. is one of the greatest of all time. It is also one of those books that folks like to say, "Yea, sure I have a copy," but have never really read, studied, or applied to operations.

Ashes to Ashes: the Phoenix Program and the Vietnam War, by Dale Andrade. It's truly a shame that contemporary practitioners of the art of war do not study the Vietnam War. The United States did a lot of things *right* during that war. It was the politicians and hippies that encouraged us to forget about our victories there.

The Clash of Civilizations and the Remaking of the World Order, by Samuel P. Huntington. This is a book on the dynamic behind the geopolitical world we live in today, rather than a military book, *per se.* Huntington was a genius and a prophet; this will open your eyes!

The Defense of Duffer's Drift, by E.D. Swinton. This classic, 72-page book, published in 1907, was the inspiration for *The Elegance of Ruthlessness.* It's often discounted by today's American generals as politically incorrect colonialist tripe. All the more reason you should buy a copy and read it!

Double Cross: The True Story of the D-Day Spies, by Ben MacIntyre. Though I believe *The Elegance of Ruthlessness* outlines the criticality of good intelligence, I do not believe I did justice to the critical aspect of deception in warfare. This is a great story of deception and what it can achieve.

Empire of the Summer Moon, by S.C. Gwynne. Want to know why the Comanche dominated the south-central plains? They were hellish savages! Want to know how the US Army beat them? We were savagely smart. Read and learn, kids.

Fighting in Hell: The German Ordeal on the Eastern Front, edited by Peter G. Tsouras. Americans may have forgotten about Vietnam, but most have never even attempted to learn about WWII on the Eastern Front. This was a totally different war from the Western Front. It was mass, maneuver, and hell on a grand scale. It is worthy of study, especially to understand contemporary German and Russian mind-sets.

Guerrilla Strategies: An Historical Anthology from the Long March to Afghanistan, edited by Gérard Chaliand. This was one of my favorite books at West Point because it detailed many of the more obscure irregular strategies, such as the Huk Guerrillas in the Philippines to the Mau Mau insurrection in Kenya.

The Landmark Tucydides, A Comprehensive Guide to the Peloponnesian War, edited by Robert B. Strassler. Okay, okay, this is one of the few very *long* books I have recommended, but he Peloponnesian War was a very *long war*! (Not unlike our long war against Islamic terrorism.)

Making Bricks Without Straw, by John Morgan Dederer. This tiny 60 pager is a comparative analysis of the tactics of Nathanael Green's Southern Campaign and Mao Tse-Tung's Mobile War. Too many have forgotten that Americans practiced guerrilla warfare before Mao!

On Guerrilla Warfare, by Mao Tse Tung. The Chinese have been some of the most brilliant military theorists in history. Failure to study their methods and theories is to set yourself up for failure.

On War, by Carl von Clausewitz. One of the greatest books on the art of war of all time, this ranks up there with Sun Tzu's *Art of War* and Thomas St. James' **Elegance of Ruthlessness* (*just kidding!!). Serious students of the art of war will read and study Clausewitz thoroughly. If you're too lazy to read the whole thing, then start with Michael Howard's *Clausewitz: A Very Short Introduction.*

One Tribe at a Time: A Strategy for Success in Afghanistan, by Major Jim Gant. Major Gant, a highly accomplished green beret, was known as someone who kind of "went native" during his many tours in Afghanistan. Ah! Know thy enemy & learnest how to vanquish him! This short book is only about 40 pages and free to download off the internet.

The Origins of War, by Donald Kagan. Kagan is a modern genius in the study of the origins and evolution of warfare and regarded as the modern torchbearer for *Realpolitik.*

Partisans and Redcoats, by Walter Edgar. This is another small book on the brilliance and unorthodox methods used to grind down the British Army in the South during the American Revolutionary War.

The Prince and Other Writings by Niccolo Machiavelli, translated by Wayne A. Rebhorn. Machiavelli has, unfortunately, become a contemporary pejorative. Looking about the geopolitical landscape of the early 21st Century, I don't understand why. His ideas are practiced all the time. Remember, *warfare is an extension of politics*!

Spec Ops, Case Studies in Special Operations Warfare: Theory and Practice, by Admiral William H. McRaven. McRaven (another great Texas hero), was humble in not including his own book in the list recommended by Special Operations Command. But, this is a brilliant book! Read it and you'll see the mind behind the Abbottabad Raid that killed Osama bin Laden!

Targeted Killing: Self-Defense, Peemption, and the War on Terrorism, by Thomas B. Hunter. Hunter, an intelligence analyst with the Defense Intelligence Agency, probably didn't make a dime off this little 74-page book. That's a shame; its brilliant.

Unrestricted Warfare, by Colonels Qiang Liang and Wang Xiangsui. If you haven't read this modern Chinese classic, you are *wrong* and fail to understand the evolution of warfare. This is truly the modern version of Sun Tzu. Read it, study it, apply it if you want to win. You have no excuse - you can download it for free on the internet!

War in European History, by Michael Howard. To understand how war has evolved, you need to learn a bit about military history. This short, 144 page book is a great foundation for such.

US Special Operations Command, Commander's Recommended Reading List for 2014

<u>Strategic Context</u>

"Asymmetric Warfare: Today's Challenge to US Military Power Issues in Twenty-First Century Warfare" by Roger W. Barnett (2003)

"Afghanistan the Bear Trap: The Defeat of a Superpower" by Mohammed Yousaf and Mark Adkin (2001)

"Blink: The Power of Thinking Without Thinking" by Malcom Gladwell (2005)

"Carnage and Culture: Landmark Battles in the Rise to Western Power" by Victor Hanson (2002)

"Children of the Jihad: A Young American's Travels Among the Youth of the Middle East" by Jared Cohen (2008)

"Thinking in Time: The Uses of History for Decision Makers" by Earnest May and Richard Neustadt (1988)

"The World is Flat 3.0: A Brief History of the Twenty-First Century" by Thomas Friedman (2007)

Leadership

"Eleven Rings: The Soul of Success" by Phil Jackson (2013)

"Achilles in Vietnam: Combat Trauma and the Undoing of Character" by Jonathan Shay (1994)

"Made to Stick: Why Some Ideas Survive and Other Die" by Chip and Dan Heath (2007)

"My Share of the Task: A Memoir" by Gen. Stanley McChrystal (2013)

Military Science

"One Hundred Victories: Special Ops and the Future of American Warfare" by Linda Robinson (2013)

"Getting More: How You Can Negotiate to Succeed in Work and Life" by Stuart Diamond (2010)

Intel and Security Affairs

“Phoenix and the Birds of Prey: The CIA’s Secret Campaign to Destroy the Viet Cong” by Mark Moyar (2007)

“Architect of Global Jihad: The Life of Al-Qaida Strategist Abu Mus’ab al-Suri” by Lia’s Brynjar (2008)

The principles of War _remain_: MOM USE SOS

- **Mass**
- **Offensive**
- **Maneuver**

- **Unity of Command**
- **Simplicity**
- **Economy of Force**

- **Security**
- **Objective**
- **Surprise**

Made in the USA
Monee, IL
04 November 2026

67383571R10079